The Economic Consequences of Outdoor Air Pollution

OECD
BETTER POLICIES FOR BETTER LIVES

Please cite this publication as:
OECD (2016), *The Economic Consequences of Outdoor Air Pollution*, OECD Publishing, Paris.
http://dx.doi.org/10.1787/9789264257474-en

ISBN 978-92-64-25746-7 (print)
ISBN 978-92-64-25747-4 (PDF)

Photo credits: Cover © "Cairo Skyline" by Sebastian Horndasch, 2014 (CC BY 2.0 https://creativecommons.org/licenses/by/2.0/)

Corrigenda to OECD publications may be found on line at: *www.oecd.org/about/publishing/corrigenda.htm*.

Preface

Air pollution is the cause of alarming numbers of premature deaths, as well as serious impacts on human health and the environment. This report sheds new light on the economic consequences of significant increases in outdoor air pollution that will occur if policies are not strengthened. Unless we clean up the air, by the middle of the century one person will die prematurely every 5 seconds from outdoor air pollution. The associated costs to society are rising rapidly. Urgent action is needed to prevent these projections from becoming a grim reality.

This report presents the first comprehensive assessment of the economic consequences of outdoor air pollution in the coming decades. First, it provides a global outlook to 2060 for the major impacts of increased air pollution on human health and agriculture: numbers of premature deaths, cases of illness and loss of agricultural yields. Second, it uses a detailed modelling framework, the OECD's ENV-Linkages model, to calculate regional and global economic costs related to those impacts that can be linked to markets, such as changes in health care expenditures, labour productivity, and agricultural production. Finally, valuation techniques are employed to assess the welfare costs of outdoor air pollution relative to the costs of premature deaths as well as pain and suffering from illness. Together, these provide a unique insight into the global and regional costs of inaction on outdoor air pollution.

Further degradation of the environment and natural capital can compromise prospects for future economic growth and human well-being. In order to assess the feedbacks from the environment on economic growth, modelling tools used for projecting future pathways of economic activity need to be able continually to assess how different environmental impacts affect various elements of the economic system. This is the ambition of the OECD's CIRCLE project. The modelling tools underlying this report contribute to this ambition by quantifying the full cycle of economy and air pollution linkages. This allows a much more elaborate quantitative assessment of the economic consequences of outdoor air pollution.

This report focuses on highlighting how outdoor air pollution may grow into an even more severe global problem unless governments act decisively. For the first time, it brings together detailed projections on the global biophysical consequences of policy inaction until 2060 with severe consequences for economies and well-being around the world. Significant uncertainties remain in the valuation of the welfare costs of premature deaths and illness, but it is certain that millions of lives will be at risk and economies will be worse off unless more ambitious policy action is taken. The OECD recognises this and urges countries to act. We need better air pollution policies for longer and healthier lives.

Angel Gurría
OECD Secretary-General

Acknowledgements

This report was prepared by Elisa Lanzi and Rob Dellink of the OECD Environment Directorate. Jean Chateau (OECD) contributed to the modelling of ENV-Linkages. Mike Holland (EMRC) and Joe Spadaro (SERC) calculated the mortality and morbidity health impacts and related healthcare and welfare costs. They also provided inputs to the text. Rita Van Dingenen (EC-JRC) provided the calculations of the concentrations of the pollutants and of the impacts on crop yields and related inputs to the text. Input data on air pollutants emission coefficients was received from Zbigniew Klimont (IIASA).

The OECD Environment Policy Committee (EPOC) was responsible for the oversight of the development of the report. In addition, the Working Party on Integrating Environmental and Economic Policies (WPIEEP) and the experts following the CIRCLE project reviewed earlier drafts.

The project was managed by Shardul Agrawala, who also provided feedback on the modelling and earlier drafts. Marie-Jeanne Gaffard provided administrative and technical support; statistical and technical assistance was provided by François Chantret. This version also benefits from comments on earlier drafts by Nils-Axel Braathen, Anthony Cox, Nathalie Delrue, Marion Devaux, Olivier Durand-Lasserve, Ivan Haščič, Ada Ignaciuk, Nick Johnstone, Alexander Mackie, Walid Oueslati, Franco Sassi, Richard Sigman, Ioannis Tikoudis and Simon Upton (all OECD), as well as Frank George (WHO), Alastair Hunt (University of Bath), Tom Kram (PBL), Frederik Neuwahl and Bert Saveyn (EC-JRC), Ståle Navrud (Norwegian University of Life Sciences), Wei-Shiuen NG (ITF), and Colin Price (Bangor University).

Table of contents

Figures

Tables

Boxes

Acronyms and abbreviations

BC	Black carbon
CBA	Cost-benefit analysis
CEA	Cost-effectiveness analysis
CES	Constant elasticity of substitution
CGE	Computable general equilibrium
CH_4	Methane
CO	Carbon monoxide
CO_2	Carbon dioxide
EC-JRC	European Commission Joint Research Centre
EU	European Union
FASST	Fast Scenario Screening Tool
GAINS	Greenhouse Gas and Air Pollution Interactions and Synergies
GBD	Global Burden of Disease
GDP	Gross domestic product
GHG	Greenhouse gas
IIASA	International Institute for Applied Systems Analysis
IPCC	Intergovernmental Panel on Climate Change
NH_3	Ammonia
NO_x	Nitrogen oxides
NO_2	Nitrogen dioxide
N_2O	Nitrous oxide
OC	Organic carbon
OECD	Organisation for Economic and Co-operation and Development
O_3	Ozone
PM	Particulate matter
PPP	Purchasing power parity
QALY	Quality adjusted life year
RCP	Representative concentration pathway

RP	Revealed preference
SO_2	Sulphur dioxide
SP	Stated preference
UN	United Nations
USD	United States dollars
VOCs	Volatile organic compounds
VOLY	Value of a life year
VSL	Value of statistical life
YOLL	Year of life lost
WHO	World Health Organisation
WTA	Willingness-to-accept
WTP	Willingness-to-pay

Executive summary

Air pollution is one of the most serious environmental risks, particularly in big cities and highly populated areas. Previous work by the World Health Organisation (WHO) and others has demonstrated the alarming consequences of outdoor and indoor air pollution for human health, especially a large number of pollution-induced premature deaths. Projections of the global economic consequences of future air pollution have however been entirely lacking.

This report provides a comprehensive assessment of the regional and global economic consequences of outdoor air pollution in the coming decades. While both outdoor and indoor air pollution are currently responsible for serious health impacts and economic consequences, economic growth in the coming decades will lead to a deterioration of outdoor air quality in particular. For this reason, the present report focuses on the future economic consequences of *outdoor* air pollution.

The report addresses the impacts of outdoor air pollution on mortality, morbidity (illness), and changes in crop yields as caused by high concentrations of particulate matter ($PM_{2.5}$) and ground level ozone. Other impacts, such as those on ecosystem services, buildings or visibility, and the direct health impacts of nitrogen dioxide (NO_2), have not been quantified owing to insufficient reliable data at the global scale. The projections of the consequences of outdoor air pollution reflect the future biophysical impacts and economic costs of air pollution in the absence of additional policies. The projections thus quantify the costs of inaction, the benchmark against which the benefits of additional policy action can be evaluated.

The analysis covers the period 2015-2060 and presents the projected economic consequences of outdoor air pollution for different types of costs. First, the market costs of outdoor air pollution, focusing on labour productivity, health care expenditures and changes in crop yields, are assessed with a multi-regional, multi-sectoral dynamic general equilibrium model. The modelling approach links economic activity to emissions of air pollutants, concentrations, biophysical impacts and feedback effects from these impacts on the economy. Second, the non-market health impacts of outdoor air pollution (mortality and morbidity) are assessed and monetised. These monetised non-market impacts do not reflect actual costs to the economy, but are obtained using results from studies that directly value the individual willingness-to-pay for reducing health risks.

Inevitably, uncertainties concerning the economic projections, the quantification of the biophysical impacts of outdoor air pollution, and the evaluation of costs mean that the results need to be interpreted with due caution. However, this report provides for the first time projections of the regional and global magnitude of the economic consequences of one of the most severe environmental challenges.

Key messages

Increasing economic activity and energy demand will lead to a significant increase in global emissions of air pollutants to 2060 in the absence of more stringent policies. This is based on projections obtained from the OECD's ENV-Linkages model, which includes pollutants such as sulphur dioxide (SO_2), nitrous oxides (NO_x), and black carbon (BC). Despite a partial decoupling of economic activity and air pollutant emissions in some areas, emissions are projected to grow particularly rapidly in regions with higher economic growth or with increasing shares of energy and energy-intensive sectors (especially coal-based electricity generation), such as South and South East Asia and the Sub-Saharan Africa regions. Thanks to more stringent policies, emissions from OECD countries tend to be stable or to decline in the short and medium run, then flatten out or increase again as the effects of the current policies fade.

Rising emissions of air pollutants are projected to lead to increasing concentrations of particulate matter ($PM_{2.5}$) and ground level ozone. In many places, concentrations of $PM_{2.5}$ and ozone are already well above the levels recommended by the WHO Air quality guidelines. Population-weighted average $PM_{2.5}$ concentrations are already high and rapidly rising in South and East Asia, especially the People's Republic of China (henceforth "China") and India. In large parts of North America, Europe and Africa, $PM_{2.5}$ concentrations from anthropogenic sources are also high but are not projected to rise as quickly. Ozone concentrations are particularly high in Korea, the Middle East and the Mediterranean, but they also exceed air quality guidelines in many other OECD and non-OECD regions. These areas are the most polluted at present and remain so in the projections for the coming decades. High average population-weighted concentrations mean that in many areas – and especially in large cities – air pollution is permanently above recommended levels; furthermore, for several days per year, they may reach levels that are extremely dangerous for human health.

The most dangerous consequences from outdoor air pollution are related to the number of premature deaths. This report projects an increase in the number of premature deaths due to outdoor air pollution from approximately 3 million people in 2010, in line with the latest Global Burden of Disease estimates, to 6-9 million annually in 2060, in the absence of more stringent policies. By 2060, a large number of deaths are projected to take place in densely populated regions with high concentrations of $PM_{2.5}$ and ozone (especially China and India) and in regions with aging populations, such as China and Eastern Europe. The projected mortality effects of $PM_{2.5}$ exposure are much larger than those of ozone.

In addition, increasing concentrations of $PM_{2.5}$ and ozone are projected to lead to substantially more cases of illness. This will imply more hospital admissions, additional health expenditures, a high number of lost working days and limitations to normal daily activities. It is projected that the air pollution-related healthcare costs will increase from USD 21 billion (using constant 2010 USD and PPP exchange rates) in 2015 to USD 176 billion in 2060, reflecting both a large number of additional cases of illness due to air pollution, and a projected increase in the healthcare costs per illness. By 2060, the annual number of lost working days, which affects labour productivity, are projected to reach 3.7 billion (currently around 1.2 billion) for the world.

The market costs of air pollution, flowing from reduced labour productivity, additional health expenditures and crop yield losses, are projected to lead to global annual economic costs of 1% of global gross domestic product (GDP) by 2060. The projected GDP losses are especially large in China (–2.6%), the Caspian region (–3.1%) and Eastern Europe

(Non-OECD EU –2.7% and Other Europe –2.0%), where air pollution impacts lead to a reduction in capital accumulation and a slowdown in economic growth.

In addition to the market costs, the report also presents projections of non-market costs associated with increased mortality and morbidity due to outdoor air pollution. These non-market costs (also referred to as welfare costs) differ from market costs in that they are based on people's expressed willingness to pay to reduce health risks and do not represent an actual cost to the economy. They provide a useful indication for policy makers of the importance of the health impacts of outdoor air pollution.

The annual global welfare costs associated with the premature deaths from outdoor air pollution, calculated using estimates of the individual willingness-to-pay to reduce the risk of premature death, are projected to rise from USD 3 trillion in 2015 to USD 18-25 trillion in 2060. In addition, the annual global welfare costs associated with pain and suffering from illness are projected to be around USD 2.2 trillion by 2060, up from around USD 300 billion in 2015, based on results from studies valuating the willingness-to-pay to reduce health risks. In per capita terms, the average global welfare costs from mortality and morbidity are projected to increase from less than USD 500 per person in 2015 to around USD 2 100-2 800 in 2060.

The potential economic consequences of both the market and non-market impacts of outdoor air pollution are very significant and underscore the need for strong policy action. Policies to limit air pollution emissions would lead to an improvement in air quality, reduce risks of very severe impacts, as well as generate considerable climate co-benefits. However, as both the sources of air pollutant emissions and the consequences of outdoor air pollution on human health and the economy are very unequally distributed across different regions, policies that reduce pollution levels and protect vulnerable groups of the population from the worst health impacts must be at the heart of an optimal policy mix.

Chapter 1

The links between outdoor air pollution and economic growth

This chapter first presents the main approaches used in the literature to assess the costs of inaction or benefits of action for air pollution. It then introduces the methodology used in this report to study the economic consequences of outdoor air pollution, using a general equilibrium model for market impacts and results of direct valuation studies for non-market impacts. The chapter also presents an overview of the main impacts of outdoor air pollution, including those related to human health and the environment. It then highlights which impacts and economic consequences are quantified in this report. The chapter ends with a description of possible policy approaches to address outdoor air pollution.

1.1. Introduction

Air pollution is one of the most serious environmental risks, particularly in big cities and highly populated areas where it causes strong negative impacts on human health. Outdoor air pollution has also been recognised to have consequences for the environment, with impacts on crop yields, biodiversity, land and water, and on human activities, with impacts on visibility and on buildings and materials, including cultural heritage.

Previous work shows alarming results on the severe impacts of outdoor and indoor air pollution on human health and in particular on the large number of premature deaths it causes.[1] The most recent Global Burden of Disease (GBD) study finds that air pollution – indoor and outdoor combined – is the top cause of environment-related deaths worldwide and estimates it was the cause of 5.5 million premature deaths globally in 2013 (Forouzanfar et al., 2015; Brauer et al., 2016). This is equivalent to 1 in 8 deaths worldwide. The 2010 GBD study (Lim et al., 2012), WHO (2014) and Lelieveld et al. (2015) estimate that *outdoor* air pollution alone is the cause of 3 to 4 million premature deaths per year at global level. According to WHO (2016), 98% of cities in low- and middle income countries and 56% of cities in high-income countries do not meet WHO air quality guidelines. The precise numbers generated by different studies are variable, reflecting refinements for example with respect to exposure modelling (e.g. the exposure cut-off point, or the slope and shape of exposure-response functions). However, the studies are consistent in showing that air pollution has a substantial effect on health and that it can be associated with several million deaths each year.

The negative impacts of air pollution on health and the environment also lead to high economic costs. OECD (2014) uses the "value of a statistical life" (VSL) to estimate the economic costs of outdoor air pollution. It finds that the cost of the health impacts of air pollution in OECD countries (including deaths and illness) was USD 1.7 trillion in 2010.[2] The cost of the health impact of air pollution in 2010 was estimated to be USD 1.4 trillion in the People's Republic of China (henceforth "China"), and USD 0.5 trillion in India.

It is less clear how the impacts and costs of air pollution will evolve in the coming decades. This report aims to fill that gap by assessing the *costs of inaction on outdoor air pollution* for a baseline projection from 2015 to 2060 at the regional and global level to 2060.[3] It focuses on the future biophysical and economic consequences of air pollution in absence of policies other than the ones that are already in place. The report shows that air pollution will have serious consequences for human health and on economic growth, unless more ambitious policies are put in place. This assessment of the costs of inaction of air pollution underlines the magnitude of the air pollution problem at global level.

The social and welfare costs of indoor air pollution should not be ignored. Indoor air pollution particularly affects poor rural communities with scarce or no access to electricity and that are affected by toxic emissions from cooking stoves, heating and lighting in their homes. Nevertheless, this report only considers outdoor air pollution. The reason is two-fold. First, the health problems associated with indoor air pollution are expected to decrease in the coming decades, even without specific new pollution control policies, as countries develop and access to cleaner energy sources becomes more widespread (cf. OECD, 2012). In contrast, the consequences of outdoor air pollution are expected to become more severe over time if no further policy actions are taken. Second, outdoor air pollution is much more directly related to economic activity, and thus a by-product of economic growth, which is the focus of this report.

The analysis in this report is based on the so-called impact pathway approach. This approach, which was developed under the EC-US Fuel Cycles Study and the ExternE project (ExternE, 1995; European Commission, 2005; US DOE, 1992), calculates the economic costs of air pollution (or the economic benefits of reduced air pollution) starting from emissions, through concentrations, exposure, biophysical impacts and valuation of the economic costs.

Previous studies have used the impact pathway approach in the context of an economic valuation of air pollution, mostly for the United States and the European Union. For the EU, such an approach was used to study the benefits of several Directives and technology options aimed to improve air quality (European Commission, 2013 and 2005; Vrontisi et al., 2016; WHO, 2013a,b; Holland, 2014a,b; ExternE, 1995; Rabl et al., 2014). For the United States, the EPA has evaluated the benefits of the Clean Air Act (US EPA, 1997, 1999, 2011). A series of studies have also been carried out on the costs of health impacts of air quality for specific regions (Matus, 2005; Matus et al., 2008; 2011; Nam et al., 2009; OECD, 2014b) and, for ozone only, at global level (Selin et al., 2009).

The report considers a set of selected impacts on health and agriculture as linked to emissions of key primary pollutants – sulphur dioxide (SO_2), nitrogen oxides (NO_x), black carbon (BC), organic carbon (OC), carbon monoxide (CO), volatile organic compounds (VOCs)[4] and ammonia (NH_3) – and the concentrations of particulate matter ($PM_{2.5}$) and ground level ozone (O_3), which are formed as a result of these emissions. Data on regional emissions for the primary pollutants was obtained from the GAINS (Greenhouse Gas and Air Pollution Interactions and Synergies) developed at the International Institute for Applied Systems Analysis (IIASA). Given the lack of reliable data at global level, it was not possible to quantify other impacts of air pollution, such as those on biodiversity or cultural heritage, or the direct impact of nitrogen dioxide (NO_2) on human health.

The analysis is based on the OECD's computable general equilibrium (CGE) model ENV-Linkages (Chateau et al., 2014). The ENV-Linkages model is used to construct a socio-economic baseline and to formulate a corresponding projection of future emissions of air pollutants. Emissions of air pollutants are then translated to concentrations of $PM_{2.5}$ and ozone using the atmospheric transportation model TM5-FASST (Fast Scenario Screening Tool) developed at the European Commission Joint Research Centre (EC-JRC). The concentration levels are the main inputs to calculate the biological and physical impacts of air pollution on human health and on crop yields. Impacts of air pollution on crop yields are calculated with TM5-FASST using the methodology of Van Dingenen et al. (2009) while a range of health impacts are calculated expanding the methodology of Holland (2014a,b) to the global level. These projections of the biophysical consequences of outdoor air pollution are then used as input to the ENV-Linkages model to calculate the projected economic costs on gross domestic product (GDP) and production.

This report presents the economic consequences of outdoor air pollution for different types of costs. While the market costs, i.e. those associated with impacts that directly affect the economy, are calculated using the ENV-Linkages model, the non-market costs are monetised using results from stated preference (SP) studies, which directly value the willingness-to-pay (WTP) for a reduction in environmental risks. Considering these two complementary aspects of the economic costs of air pollution makes the results of this report very relevant for policy makers, as both types of costs need to be considered when designing policy responses.

This report is structured as follows. Chapter 2 presents the methodology and modelling framework used for projecting and analysing the costs of inaction on air pollution. Chapter 3 presents the projections of economic growth, emissions, concentrations and biophysical impacts of air pollution. Chapter 4 presents results on the macroeconomic costs of air pollution. Finally, Chapter 5 presents the non-market costs of air pollution, including both mortality and morbidity, and a comparison of market and non-market costs.

1.2. Main consequences of outdoor air pollution

The impacts of outdoor air pollution on health and the environment are linked to high concentrations of fine and coarse shares of particulate matter (PM), ground level ozone (O_3) and other pollutants, such as nitrogen dioxide (NO_2) and sulphur dioxide (SO_2).

PM includes both primary particulates emitted in the atmosphere, such as black carbon (BC), organic carbon (OC), metals, salts and ashes, and secondary particulates, which are formed in the atmosphere from a reaction among precursor gases. The precursor gases of PM include ammonia (NH_3), nitrogen oxides (NO_x), sulphur dioxide (SO_2) and, to some extent, volatile organic compounds (VOCs). Ground level ozone is formed in the atmosphere as a consequence of chemical and photochemical reactions involving precursor gases such as NO_x, VOCs and methane (CH_4).[5]

Concentrations of the pollutants are a composite effect of emissions from anthropogenic and natural sources (dust, sea salt, volcanoes, forest fires, etc.). Some geographical areas, such as the Mediterranean Sea or areas to the south of the Sahara desert, have high levels of natural PM (sea salt and dust). Background concentrations of ozone are always present in the atmosphere, but air pollutant emissions increase concentrations regionally. Concentrations of pollutants also depend on climatic conditions. For instance, sunlight increases the presence of ozone in the atmosphere, while a lack of precipitation leads to higher concentrations of PM.

Several other factors influence concentrations and the possibilities of dispersion of the pollutants in the atmosphere. Characteristics linked to the location of the emissions, such as the volume and geographical location of emissions, the topography of the location, whether the emissions are from fixed or mobile sources, and the presence of winds affect the dispersion possibilities. Chemical characteristics of the pollutants, such as the lifetime of the pollutants in the atmosphere, and the capacity of the pollutants to convert into secondary pollutants, also affect concentrations.

A large share of primary emissions is caused by fuel combustion due to fossil-fuel based power generation, transport, industry, and burning of traditional biomass in the residential sector. Some industrial processes also cause large emissions, especially when there is an extensive use of chemical substances. Significant emissions also come from the use of fertilisers, agricultural waste, savannah burning and forest fires.

Spikes in air pollution and long-term exposure to high concentrations of air pollutants affect human health, causing increase in both mortality (i.e. the number of premature deaths attributable to air pollution) and morbidity (i.e. the increase in the incidences of illnesses due to air pollution). Pollution-related illnesses include lung cancer, cardiovascular diseases (ischemic heart disease and stroke), respiratory diseases (chronic bronchitis and asthma) and chronic obstructive pulmonary diseases (WHO, 2013b; Hunt et al., 2016). The additional cases of illness result in more hospital admissions, medical expenses and absences from work. In turn, the absences from work can lead to a reduced productivity of labour. However, air pollution can also have a direct impact on labour productivity, without resulting in absences from work (Graff-Zivin and Neidell, 2012).

An emerging literature shows that air pollution has additional health impacts on fertility, pregnancy, birth weight, and new-borns and children. Effects on new-borns and children may result in neurodevelopment and cognitive issues, which in turn can affect performance at school, and, further in life, lead to lower earnings.

High concentrations of PM, especially finer particles ($PM_{2.5}$), are the main cause of health impacts, as they can easily penetrate into the lungs and bloodstream. There are also direct health impacts due to high concentrations of other pollutants, such as ozone, SO_2 and NO_2 (see WHO, 2013b; Walton et al., 2015). A recent report by the Royal College of Physicians (RCP, 2016) provides an estimate for the combined effect of PM and NO_2 in the UK of 40 000 deaths per year (±25%), an increase from a generally accepted figure of 29 000 for $PM_{2.5}$ alone. This estimate pays particular attention to the potential for overlap in estimates of mortality from assessment of PM and NO_2 in isolation of each other.

High concentrations of ground level ozone also lead to negative impacts on crops yields, as well as plants in general. As a strong oxidant, ozone is toxic to plants and causes several types of symptoms including markings on the foliage (which can make leaf crops such as spinach or lettuce unsaleable), reduced growth and yield, as well as premature death of the plants.

Air pollution also has other negative effects on the environment, including on forests and biodiversity, water and land. It can lead to reduced visibility ("smog"), which limits vistas in national parks and protected areas, affects safety and human activities, and ecosystems. Finally, acidic and nitrogen compounds in the air can deposit onto land and water, degrading water quality and affecting ecosystems with consequences for food quality (and thus for human health), and for the commercial and recreational use of the affected areas. High nitrogen deposition is now recognised as a major threat to biodiversity and overall ecosystem health (Sutton et al., 2011).

The biophysical impacts of outdoor air pollution entail large economic costs. Impacts on human health dominate the "costs of inaction" on air pollution, representing about 90% of the total social costs for some pollutants (OECD, 2008). The health impacts of air pollution lead to increased health expenditures as well as labour productivity losses. Reduced agricultural output can also cause economic losses especially in areas where agriculture constitutes a large part of the economy. Finally, high concentrations of air pollutants, reduced visibility and damages to buildings and cultural heritage can all have consequences for tourism and hence economic costs due to reduced tourism flows.

While this report only focuses on outdoor air pollution, indoor air pollution also poses serious risks to human health. WHO (2014) estimated 3.7 million deaths attributable to outdoor air pollution, but as many as 4.3 million deaths to indoor air pollution.[6] The most significant source is burning of traditional solid fuels such as coal and biomass (e.g. cow dung and wood) for indoor cooking and heating by households, which cannot afford cleaner fuels. Indoor air pollution is also a concern in developed countries, mainly from releases of chemicals from carpets, furniture and household cleaning products, as well as radon and pesticides. OECD (2012) provides some comparisons of the health effects of indoor compared with outdoor pollution.[7] They find that with raising income levels in emerging and developing countries and improved access to commercial energy sources and to health services, indoor air pollution will gradually become less important in comparison with outdoor air pollution.

1.3. Selected impacts of outdoor air pollution

Ideally the analysis in this report would cover all the impacts and costs of outdoor air pollution described in the previous section. Owing to a lack of available data however, it was only possible to assess the costs of air pollution of a selected number of impacts, which are deemed to be of high importance. This report considers impacts of $PM_{2.5}$ and ground level ozone on human health and on agricultural crop yields, as summarised in Table 1.1.

More specifically, the health impacts considered are premature deaths and increasing cases of illnesses (cardiovascular and respiratory diseases). The market-related impacts modelled in ENV-Linkages are thus increased health expenditures, reduced labour productivity as linked to absences from work due to illness and reduced crop yields, while non-market costs related to mortality and morbidity are calculated separately using results from SP studies.

Table 1.1. **Main outdoor air pollution impact categories**

Impact category	Impacts description	Market impacts	Non-market impacts
Health	Mortality from lung cancer, cardiovascular and respiratory diseases due to high concentrations of $PM_{2.5}$ and ozone		Premature deaths
	Morbidity from lung cancer, cardiovascular and respiratory diseases due to high concentrations of $PM_{2.5}$ and ozone	Increased health expenditures Changes in labour productivity due to absence from work for illness	Disutility (e.g. pain and suffering) due to illness
	Other health impacts, from e.g. low birth weight, pregnancy	Not covered in this report	
	Direct health impacts from NO_2	Not covered in this report	
Agriculture	Damages to crop yields due to high concentrations of ozone	Changes in crop yields	
Tourism, leisure	Changes in tourism and leisure due to e.g. reduced visibility, damages to cultural heritage and health risks	Not covered in this report	
Ecosystems, biodiversity, forestry	Degraded air and water quality, reduced ecosystem health	Not covered in this report	

Impacts on cultural heritage, tourism, leisure activities, forestry and biodiversity could not be included as there is not yet enough information available either to attribute the impacts on air pollution or to quantify the impacts in monetary terms.[8] Some health impacts, such as those on pregnancy and birth weight and the direct effects of NO_2 exposure, were also omitted owing to lack of information.

This report can also only account for a subset of all economic consequences originating from the impacts considered. One prime example of an effect that cannot be included in the modelling framework is the (indirect) economic consequences of premature deaths on labour markets. In principle, labour supply lowers if a person from the working age population dies. Similarly, a premature death can also affect future labour supply through a reduced number of births and hence a decrease in the population. But such endogenous effects are not easily predictable and beyond the scope of the current report. Further, for outdoor air pollution these effects can be expected to be relatively small as the premature deaths mostly concern elderly people, with no effects on the labour force and on future population growth (see also Section 4.3). Therefore, the assumption is taken that mortality does not affect the labour market.

As already discussed, the analysis is limited to outdoor air pollution only and does not consider indoor air pollution. Unfortunately, there is very little literature on the economic consequences of indoor air pollution, especially on those related to "new" chemical sources of pollution, so a robust quantitative assessment is not yet possible.

With these missing elements, it is likely that the present report underestimates the costs of air pollution. There are also major uncertainties in the analysis, particularly those involving making projections for future behaviour. Uncertainties exist in the socioeconomic projections, in the projections on the future structure of the economy, via emissions and concentrations of air pollutants to the health impacts and the effects thereof on the economy. While there are important uncertainties at every stage, there is no robust literature to assess which uncertainties are most important for the conclusions of this report. Therefore, the reader should keep in mind the presence of uncertainties throughout the report and in the results presented.

The uncertainties involved in quantitative studies should not unduly deter action, as a number of qualitative insights are robust, most importantly that outdoor air pollution affects health, as demonstrated by repeated epidemiological studies undertaken throughout the world, and that associated impacts on the economy and on welfare are substantial.

1.4. Typology of air pollution costs

This report considers both market and non-market costs of outdoor air pollution. Market costs (i.e. costs to the economy) are those that are associated with biophysical impacts that directly affect economic activity as measured in the national accounts and GDP. For example, lower crop yields affect agricultural production. Non-market costs include the monetised welfare costs of mortality (premature deaths), and of the disutility of illness (e.g. pain and suffering). While market costs show the need to address air pollution policies in order to avoid negative effects on the economy, non-market costs show the – potentially extremely high – social benefits that air pollution control policies can have. Figure 1.1 graphically represents the different types of costs considered in this report.

Figure 1.1. **Cost categories considered in this report**

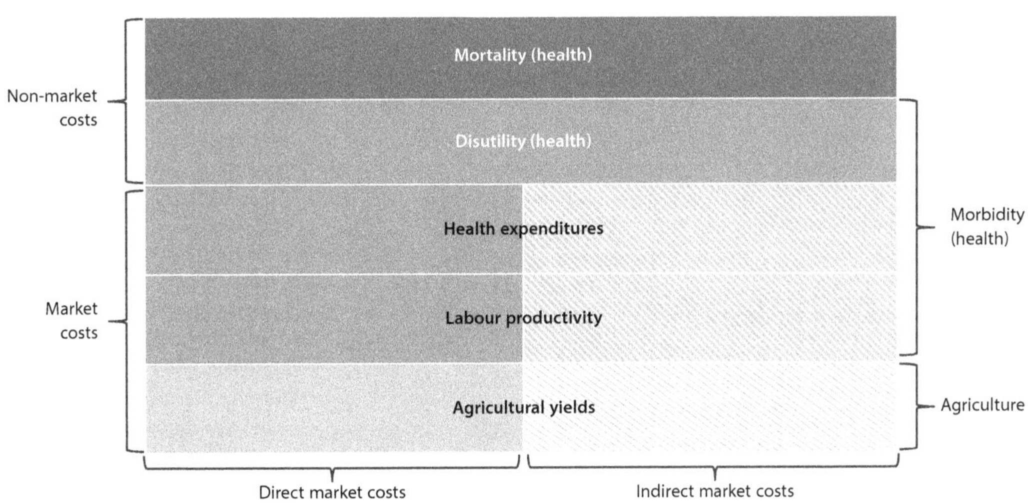

The market-related impacts, which in this study comprise additional health expenditures due to illness, labour productivity losses due to absences from work for illness, and agricultural yield losses, are included in the ENV-Linkages model to calculate the global and regional costs of outdoor air pollution on GDP and sectoral production. The market costs are further split into direct and indirect market costs. A general equilibrium framework can take into consideration both direct and indirect effects throughout the economy. For instance, a decrease in crop yields will lead to a direct impact on agricultural output of the affected crops, but also to indirect effects, including substitution by other crops and changes in trade patterns. As underlined by Hunt et al. (2016), since the market impacts of air pollution may result in significant effects on related markets or government finances, an economy-wide modelling approach is needed to capture the full economic costs.

Non-market impacts cannot be easily accounted for in a general equilibrium framework as they are not linked to any specific variable in the production or utility functions of the model. The welfare costs of non-market impacts, including both the costs of mortality and morbidity caused by outdoor air pollution, are evaluated using results from SP studies.[9]

To compare market and non-market costs, both types of costs can be expressed in terms of welfare. Non-market costs are directly calculated as welfare costs. Market costs can be expressed as welfare costs using the concept of equivalent variation of income.[10] This, as well as the comparability of the different types of costs, is further discussed in Sections 2.8 and 5.4.

1.5. Possible policy responses to outdoor air pollution

This report only focuses on the costs of inaction of outdoor air pollution. Nevertheless, there are several policy options available to address air pollution. A taxonomy of policy instruments to address air pollution is summarised in Table 1.2, which is reproduced from OECD (2012).

The implementation of policies that reduce pollution levels will certainly address and reduce the biophysical as well as the economic costs of air pollution. These can include incentivising or requiring the adoption of end-of-pipe technologies that can reduce pollution or of cleaner technologies, especially for energy combustion, as well as implementing air quality standards, automobile emission standards, fuel quality standards, and emission taxes, among others.

Table 1.2. **Taxonomy of policy approaches for air pollution management**

<table>
<tr><th>Regulatory approaches</th><th>Economic instruments</th><th>Others</th></tr>
<tr><td>• Ambient air quality standards.
• Emission ceilings (e.g. the European Union's National Emission Ceiling Directive).
• Industrial emission standards, technology standards.
• Reporting requirements for stationary sources (e.g. pollutant release and transfer registers).
• Fuel efficiency standards.
• Fuel quality standards.
• Vehicle inspection and maintenance programmes.</td><td>• Tradable permits schemes for air emissions from stationary sources (e.g. SO_2 allowance trading system under the US Clean Air Act).
• Fuel taxes.
• Road pricing.
• Congestion charges.
• Taxes on emissions.
• Financial incentives for the development of alternative and renewable fuels and advanced transport technologies (e.g. California's DRIVE programme).</td><td>• Information collection:
- through emission and air quality monitoring;
- for cost-benefit analyses to support policy evaluation (with valuation of health impacts);
- for public education (e.g. Canada's Air Quality Health Index).
• Voluntary schemes (e.g. car-scrapping schemes).
• International conventions (e.g. The Convention on Long-range Transboundary Air Pollution).
• Infrastructures and urban planning.
• Flexible work initiatives (e.g. the US Telework Enhancement Act of 2010).</td></tr>
</table>

Source: OECD (2012).

Education, information diffusion, cohesion policies and early warnings can also reduce the impacts of air pollution on health. Cohesion policies can provide support for countries to comply with legislations, develop infrastructures and respond to environmental challenges with improved organisational resources. Warning the population of spikes of air pollution and restricting activities, especially for the populations at higher risk, can reduce the health impacts. However, this may also require more flexibility in terms of working hours or telework initiatives, if possible, in order to avoid high impacts on the labour market. The efficiency of flexible work initiatives depends on the stage of economic development and it may be beneficial only in countries with a high share of services sectors (rather than e.g. industrial).

Human exposure to air pollution has a spatial dimension. This is because both population density and the resulting pollutant concentrations vary over space. This creates a role for both local initiatives and measures that do not take specific account of local factors, such as vehicle or industrial emission standards. Effective local policies, aiming at reducing pollution levels in highly populated areas include industrial relocation, spatially-differentiated pollution taxes and environmental and residential zoning (Cárdenas Rodríguez et al., 2015). Moreover, lower income groups are usually more exposed to pollution, as they are often located in more polluted and populated areas (where housing costs are lower). They also usually have longer commutes with exposure to high concentrations on roadways, and have (in many cases) restricted access to healthcare. Therefore, spatial considerations need to be recognised when designing air pollution control policies.

Even if air pollution mostly has local and regional consequences, it is also a global problem. Several pollutants and small particles such as PM can be transported by winds and have impacts in regions and countries other than the ones where they have been emitted. Further, air quality is deteriorated in almost all major regions of the world, and international linkages between countries, not least through international trade, mean that changes in consumption patterns in one country affect emission levels in others. The high pollution levels in China are not only a consequence of increasing domestic consumption, but also of production activities for export purposes. Global solutions are also needed to develop less polluting technologies, and a global transformation of the energy system is an essential part of any cost-effective policy response (IEA, 2016).

Many countries are actively taking steps to avoid the direst consequences of inaction of air pollution. If these policy plans are effectively implemented and followed by more ambitious policies, the costs of inaction as portrayed in this report will not materialise in full. As discussed in Chapter 2, this report does not provide a prediction of what will happen, but a plausible projection of what might happen if countries do not undertake any further efforts to reduce emissions below the levels that result from current legislation.

Further, there are strong interactions with a wide variety of other policy domains. Reducing air pollution provides an opportunity to reap synergies with investments in green growth, green technology, green infrastructure, and with promoting innovation. An overarching sustainable development framework that encompasses a country-specific sustainable development strategy and that promotes green growth, clean technologies, and less inequality and poverty would provide an integrated policy response that would include the multiple benefits of co-ordinated action. Such an integrated policy response can help exploit synergies between different policy objectives and avoid harmful contradictions between uncoordinated regulations.

There are strong interactions between air quality measures and climate or energy policies. A cleaner energy sector or the implementation of climate policies will also lead to lower emissions of air pollutants as well as higher cost-efficiency. It is therefore important to stress the need for integrated policies that consider trade-offs and co-benefits for policy objectives on climate change, energy and air pollution. Stimulating energy efficiency is the typical example of an integrated policy response that has multiple benefits (IEA, 2014).

The consequences of air pollution also have strong interactions with health care policy implementation. For instance, the improved availability and effectiveness of health infrastructure can help reduce the negative impacts on both labour productivity and the disutility of illness. With air pollution worsening in many parts of the world, there may also be more research, which will lead to a better understanding of exposure to high levels of concentrations of air pollutants and to better understanding of the burden of other diseases. The availability of this type of information will also help find responses in terms of cures and recommendations.

There is no one-fits-all recipe for reducing the impacts of air pollution. There are large differences among countries in terms of prevalent pollutants and sources. In general, a mix of policy instruments provides flexibility and wide coverage, although undue overlap between policy instruments should be avoided. Analysing the sources and causes of emissions in each country can guide towards the choice of the optimal policy mix and avoid policies in one sector harming another. A co-ordinated policy mix among different environmental issues is essential. This would avoid policy trade-offs such as achieving renewable energy targets by increasing biomass use for heating, while causing an increase in local PM pollution.

Notes

1. A death can be classified as premature if it happens before the expected age, as related to the life expectancy of a country, gender or specific health state, and if it can be prevented by reducing the cause of death, in this case outdoor air pollution.

2. Throughout this report, monetary values are presented in constant 2010 US dollars using purchasing power parity (PPP) exchange rates (i.e. "international dollars"), unless otherwise indicated. For brevity, this is indicated in the text simply as "USD".

3. The term "region" is applied loosely throughout the report; ENV-Linkages contains a combination of 12 major countries and 13 groups of countries. These are generically referred to as regions.

4. VOCs refer to a group of carbon-based chemicals (such as acetone, benzene, formaldehyde and toluene). Each and every chemical has its own toxicity and different effects on human health. In principle, methane (CH_4) is also a VOC, but it is considered separately here, so the group of VOCs referred to throughout this report excludes methane. This group of pollutants is often referred to as non-methane VOCs (NMVOCs).

5. Ozone occurs in significant quantities both as a pollutant and as a natural component of the atmosphere. At higher elevations, ozone screens out harmful ultraviolet radiation. However, close to the ground, ozone is harmful to human health, vegetation and some materials. There is a complex relationship between ozone and nitrogen oxides (NO_x); under some conditions, emissions of NO_x will lead to ozone formation, while under others they will lead to a reduction in local ozone levels.

6. Many people are exposed to both indoor and outdoor air pollution. Thus, mortality attributed to the two sources cannot simply be added together. The total estimate of WHO is around

7 million deaths in 2012. Further, the Global Burden of Disease initiative, using slightly different response functions and exposure estimates, obtains a lower figure for outdoor air pollution mortality of 3.1 million deaths per year for 2013 from exposure to PM and ozone.

7. This report also identifies a number of ways in which climate change policies may worsen indoor air pollution, not least through higher fuel prices (which may drive poor households back to traditional biomass use) and improved insulation.

8. Some literature exists that attempts to calculate the total value of specific activities and heritage sites, not least the Economics of Ecosystems and Biodiversity (TEEB) study and the UK National Ecosystem Assessment, but these cannot be used to value the associated impacts of outdoor air pollution, not least because of problems of attribution.

9. Willingness-to-pay (WTP) measures how much money (income) a person is willing to pay to avoid or reduce the risk of a negative outcome to materialise. In the current context, it aims to measure how much people are willing to pay to avoid an increase in their risk of dying prematurely or falling ill because of outdoor air pollution. WTP is often measured through stated preference methods, i.e. questionnaires where respondents indicate their WTP value.

10. The difference in methodologies for the estimation complicates the comparability of market and non-market costs.

References

Brauer, M. et al. (2016), "Ambient Air Pollution Exposure Estimation for the Global Burden of Disease 2013", *Environmental Science and Technology*, Vol. 50, pp. 79-88.

Cárdenas Rodríguez, M., L. Dupont-Courtade and W. Oueslati (2015), "Air Pollution and Urban Structure Linkages: Evidence from European Cities", *OECD Environment Working Papers*, No. 96, OECD Publishing, Paris, http://dx.doi.org/10.1787/5jrp6w9xlbq6-en.

Chateau, J., R. Dellink and E. Lanzi (2014), "An Overview of the OECD ENV-Linkages Model: Version 3", *OECD Environment Working Papers*, No. 65, OECD Publishing, Paris, http://dx.doi.org/10.1787/5jz2qck2b2vd-en.

European Commission (2013), "The Clean Air Policy Package", *Commission Staff Working Document. Impact Assessment,* No. SWD(2013)531, European Commission, Brussels, http://ec.europa.eu/governance/impact/ia_carried_out/docs/ia_2013/swd_2013_0531_en.pdf (last accessed on 3 May 2016).

European Commission (2005), "ExternE: Externalities of Energy, Methodology, 2005 update", No. EUR 21951, Office for Official Publications of the European Communities, Luxembourg. www.externe.info/externe_d7/?q=node/30 (last accessed on 3 May 2016).

ExternE (1995), Volume 2: methodology. ExternE: Externalities of Energy. www.externe.info/externe_d7/?q=node/37 (last accessed on 3 May 2016).

Forouzanfar, M.H. et al. (2015), "Global, regional, and national comparative risk assessment of 79 behavioural, environmental and occupational, and metabolic risks or clusters of risks in 188 countries, 1990-2013: a systematic analysis for the Global Burden of Disease Study 2013", *The Lancet*, Vol. 386(10010), pp. 2287-2323.

Graff-Zivin, J. and M. Neidell (2012), "The impact of pollution on worker productivity", *American Economic Review*, Vol. 102(7), pp. 3652-3673.

Holland, M. (2014a), Cost-benefit analysis of final policy scenarios for the EU Clean Air Package. Corresponding to IIASA TSAP report no.11, Version 2a. Report to European Commission DG Environment. March 2014, http://ec.europa.eu/environment/air/pdf/TSAP%20CBA.pdf (last accessed on 3 May 2016).

Holland, M. (2014b), Implementation of the HRAPIE Recommendations for European Air Pollution CBA work. Report to European Commission DG Environment. January 2014. http://ec.europa.eu/environment/air/pdf/CBA%20HRAPIE%20implement.pdf (last accessed on 3 May 2016).

Hunt, A. et al. (2016), "Social costs of morbidity impacts of air pollution", *OECD Environment Working Papers* No. 99, OECD Publishing, Paris, http://dx.doi.org/10.1787/5jm55j7cq0lv-en.

IEA (International Energy Agency) (2016, forthcoming), *World Energy Outlook special report on Energy and Air Pollution*, IEA, Paris.

IEA (2014), *Capturing the Multiple Benefits of Energy Efficiency: A Guide to Quantifying the Value Added*, IEA, Paris, http://dx.doi.org/10.1787/9789264220720-en.

Lelieveld, J. et al. (2015), "The contribution of outdoor air pollution sources to premature mortality on a global scale", *Nature*, Vol. 525, pp. 367-371.

Lim, S.S. et al. (2012), "A comparative risk assessment of burden of disease and injury attributable to 67 risk factors and risk factor clusters in 21 regions, 1990-2010: A systematic analysis for the Global Burden of Disease Study 2010", *The Lancet,* Vol. 380, pp. 2224-60.

Matus, K. (2005), "Health Impacts from Urban Air Pollution in China: The Burden to the Economy and the Benefits of Policy", *Master of Science Thesis in Technology and Policy*, Massachusetts Institute of Technology, Cambridge, MA.

Matus, K. et al. (2011), "Health Damages from Air Pollution in China", *MIT Joint Program on the Science and Policy of Global Change Report Series*, Report No. 196, Massachusetts Institute of Technology, Cambridge, MA.

Matus, K. et al. (2008), "Toward integrated assessment of environmental change: air pollution health effects in the USA", *Climatic Change*, Vol. 88, pp. 59-92.

Nam, K.-M. et al. (2009), "Measuring welfare loss caused by air pollution in Europe: a CGE analysis", *MIT Joint Program on the Science and Policy of Global Change Report Series*, Report 178, Massachusetts Institute of Technology, Cambridge, MA.

OECD (2014), *The Cost of Air Pollution: Health Impacts of Road Transport*, OECD Publishing, Paris, http://dx.doi.org/10.1787/9789264210448-en.

OECD (2012), *OECD Environmental Outlook to 2050: The Consequences of Inaction*, OECD Publishing, Paris, http://dx.doi.org/10.1787/9789264122246-en.

OECD (2008), *Costs of Inaction on Key Environmental Challenges*, OECD Publishing, Paris, http://dx.doi.org/10.1787/9789264045828-en.

Rabl, A. et al. (2014), *How Much is Clear Air Worth: Calculating the Benefits of Pollution Control*, Cambridge University Press, July 2014.

RCP (Royal College of Physicians) (2016), *Every breath we take: the lifelong impact of air pollution*. Report of a working party of the Royal College of Physicians. London. www.rcplondon.ac.uk/projects/outputs/every-breath-we-take-lifelong-impact-air-pollution (last accessed on 3 May 2016).

Selin, N.E. et al. (2009), "Global health and economic impacts of future ozone pollution", *MIT Joint Program on the Science and Policy of Global Change Report Series,* Report 177, Massachusetts Institute of Technology, Cambridge, MA.

Sutton, M.A. et al. (2011), *The European Nitrogen Assessment: Sources, Effects and Policy Perspectives.* Cambridge University Press, New York.

US DOE (United States Department of Energy) (1992), "US-EC Fuel Cycles Study: Background Document to the Approach and Issues". Report ORNL-M-2500, Oak Ridge National Laboratory, USA.

US EPA (United States Environmental Protection Agency) (2011), "Prospective Study: The Benefits and Costs of the Clean Air Act, 1990 to 2020". United States Environmental Protection Agency.

US EPA (1999), "Prospective Study: The Benefits and Costs of the Clean Air Act, 1990 to 2010". United States Environmental Protection Agency.

US EPA (1997), "Retrospective Study: The Benefits and Costs of the Clean Air Act, 1970 to 1990". United States Environmental Protection Agency.

Van Dingenen, R. et al. (2009), "The global impact of ozone on agricultural crop yields under current and future air quality legislation", *Atmospheric Environment,* Vol. 43(3), pp. 604-618.

Vrontisi, Z. et al. (2016), "Economic impacts of EU clean air policies assessed in a CGE framework", *Environmental Science and Policy* 55, pp. 54-64.

Walton, H. et al. (2015), *Understanding the Health Impacts of Air Pollution in London.* King's College London. www.london.gov.uk/sites/default/files/HIAinLondon_KingsReport_14072015_final_0.pdf (last accessed on 3 May 2016).

WHO (World Health Organisation) (2016), *Global Urban Ambient Air Pollution Database (update 2016).* The World Health Organization, Geneva, 2016.

WHO (2014), *Burden of disease from Household Air Pollution for 2012.* The World Health Organization, Geneva, 2014.

WHO (2013a), "REVIHAAP: Review of evidence on health aspects of air pollution – REVIHAAP project: final technical report". World Health Organization, Regional Office for Europe, Bonn, Germany. www.euro.who.int/en/health-topics/environment-and-health/air-quality/publications/2013/review-of-evidence-on-health-aspects-of-air-pollution-revihaap-project-final-technical-report (last accessed on 3 May 2016).

WHO (2013b), *Health risks of air pollution in Europe – HRAPIE project. Recommendations for concentration-response functions for cost-benefit analysis of particulate matter, ozone and nitrogen dioxide*, World Health Organization, Regional Office for Europe, Bonn, Germany. www.euro.who.int/en/health-topics/environment-and-health/air-quality/publications/2013/health-risks-of-air-pollution-in-europe-hrapie-project-recommendations-for-concentrationresponse-functions-for-costbenefit-analysis-of-particulate-matter,-ozone-and-nitrogen-dioxide (last accessed on 3 May 2016).

Chapter 2

A framework for assessing the economic consequences of outdoor air pollution

This chapter presents the methodology used in this report to analyse the economic consequences of outdoor air pollution. The methodology is based on the impact pathway approach, which requires multiple steps, from creating projections of air pollutant emissions, to calculating concentrations of key pollutants, calculating the biophysical impacts on health and crop yields, and calculating the economic costs with the ENV-Linkages model for market impacts and with results of direct valuation studies for non-market impacts. For each step the modelling framework and economic techniques used are explained.

2.1. Overview of the assessment framework

The framework to assess the economic consequences of outdoor air pollution links projections of economic activity to changes in air quality and to the associated biophysical and economic consequences. Modelling and projecting these consequences is done using the impact pathway approach, which requires multiple steps and the use of different techniques and modelling frameworks. Figure 2.1 summarises the different steps employed in this analysis.

First, an economic modelling framework is needed to obtain projections of economic activity, as well as the emissions that they imply. A computable general equilibrium (CGE) model, such as the OECD's ENV-Linkages model, is the ideal framework as it also includes projections of sectoral and regional economic activities. As explained in Section 2.2, the projections of economic activity to 2060 at the sectoral and regional level rely on a range of important drivers and exogenous trends, including those for demographic developments and technological change.

Second, for each year, emissions of a range of air pollutants are linked to the different economic activities as projected in step 1. In some cases, emissions are directly linked to a specific element in the production process, such as the combustion of fossil fuels. In other cases, emissions are linked to the scale of activity, and thus to production volumes. Some emissions that are not directly linked to economic activity are projected using exogenous trends. Together, these establish projections for regional emission levels, as described in Section 2.3.

Third, emissions of air pollutants are used to calculate concentrations of $PM_{2.5}$ and ozone. This step relies on an atmospheric dispersion model and on downscaling national emissions to a spatial grid of local emission levels. It delivers a "gridded map" of concentrations for the period between 2010 and 2060, which forms the basis for the assessment of the health and environmental impacts. Section 2.4 explains this step and the modelling framework used in detail.

Figure 2.1. **Steps to study the economic consequences of outdoor air pollution**

1 PROJECTIONS OF SECTORAL ACTIVITIES

- The OECD's ENV-Linkages model is used to obtain projections of sectoral activities.

2 PROJECTIONS OF AIR POLLUTANT EMISSIONS

- Data on emissions of air pollutants are linked to sectoral activities within ENV-Linkages to obtain projections of future emissions levels that are consistent with the economic baseline.

3 CONCENTRATIONS OF AIR POLLUTANTS

- The European Commission Joint Research Centre (EC-JRC)'s TM5-FASST model is used calculate concentrations of particulate matter and ozone.

4 BIOPHYSICAL IMPACTS OF OUTDOOR AIR POLLUTION

- The concentrations of PM2.5 and ozone are used to calculate the biophysical impacts on health (e.g. number of lost working days) and agriculture (e.g. crop yield changes).

5 VALUATION OF HEALTH IMPACTS OF OUTDOOR AIR POLLUTION

- Unit values are established for each health endpoint and health market impacts are evaluated and converted to economic costs (e.g. health expenditures).

6 MARKET COSTS OF OUTDOOR AIR POLLUTION

- The indicators relative to the market impacts are used to calculate the macroeconomic costs of air pollution in the ENV-Linkages model, so as to obtain an overall effect on economic growth.

7 WELFARE COSTS OF OUTDOOR AIR POLLUTION

- Impacts that cannot be included in the model are evaluated using results from direct valuation studies. Market costs are also expressed as welfare costs.

Fourth, the biological and physical impacts caused by the high levels of population-weighted concentrations of $PM_{2.5}$ and ozone are calculated using data on population, exposure to the pollutants and results of studies calibrating concentration-response functions (see Section 2.5). This step aggregates the detailed spatial concentration information to the national level, covering 181 countries for $PM_{2.5}$, and 161 for ozone. A range of indicators is used to present the biophysical impacts, to allow differentiated effects on e.g. number of lost working days, hospital admissions and agricultural productivity impacts.

Fifth, the direct economic consequences of the health impacts are calculated at the country level. This step comprises the calculation of unit values for the evaluation of the health impacts for each endpoint. For example, hospital admissions are translated into health expenditures and a welfare cost is established for each premature death. This step is further discussed in Section 2.6.

Sixth, market costs are analysed using the ENV-Linkages general equilibrium model, which is also employed in steps 1 and 2. The direct impacts in terms of agricultural yield shocks, changes in health expenditures and labour productivity changes are aggregated to the regional aggregation level of the CGE model and used as an input to calculate the economic consequences of outdoor air pollution (see Section 2.7 for more details). This step reflects the feedback of outdoor air pollution impacts on the economy, and represents the core of the assessment of the economic consequences of outdoor air pollution.

Finally, in the last step, laid out in Section 2.8, the costs that are not directly linked to any economic variable are quantified. These non-market costs are evaluated in terms of welfare changes using results from direct valuation studies.

The impacts for which there was enough reliable data for quantification and which are included in the modelling framework are those related to change in healthcare expenditures, labour productivity changes linked to lost working days, and agricultural crop yield changes. It was not possible to include other impacts, such as those on forestry, biodiversity or cultural heritage, in the modelling framework because there are no robust studies that quantify the pollution-attributable costs at the global scale.

In line with the OECD's analysis of the economic consequences of climate change (OECD, 2015), these are introduced in the model following a production function approach (for a general framework see Sue Wing and Fisher-Vanden, 2013; for an overview of modelling applications to climate change see Sue Wing and Lanzi, 2014; Vrontisi et al., 2016, use the same approach for the assessment of the EU's Clean Air Policy Package). This means that each impact is linked to variables that are at the core of the production functions underlying the model structure.

The results are presented in the form of a stream of future costs of inaction on outdoor air pollution. For a cost-benefit analysis of specific policies, the net present value of both the costs and benefits of the policy action would need to be quantified. This additional step, which is not included in this report, crucially depends on the choice of a discount rate to evaluate intertemporal changes. By presenting the economic consequences in this report as they emerge over time, rather than converted to a present value, this controversial step is avoided.

Theoretically, one could expand the modelling framework with a utility function that includes health and other relevant factors. This approach has been experimented with in e.g. Mayeres and Van Regemorter (2008), but such an approach requires very bold assumptions on the substitutability of consumption and health impacts, and is limited to morbidity impacts. Further, it is virtually impossible to find robust estimates of the substitution elasticities between these various elements in the expanded utility function for all regions. Therefore, non-market impacts, such as the economic value of premature deaths or the disutility linked to illness, are assessed outside the general equilibrium modelling framework.

2.2. Socio-economic trends in a baseline projection

The OECD's multi-region, multi-sector dynamic CGE model ENV-Linkages (see Chateau et al., 2014 and Annex A for further details on the model) is used to create a socio-economic baseline projection of sectoral and regional economic activities until 2060. The baseline projection used in this report excludes new policies and feedbacks from air pollution and climate change impacts on the economy. It serves as a reference to calculate the future costs of air pollution. The baseline projection used in this report is identical to the no-damage baseline used for the assessment of the economic consequences of climate change (OECD, 2015).

Two different baseline projections are presented in this report. The "central" projection describes a baseline projection that considers the feedback effects of outdoor air pollution on the economy. It describes the main socioeconomic trends, emissions and concentrations of air pollutants and the resulting impacts on health and agriculture. It also contains the feedbacks of these impacts on the economy. This central projection is contrasted with a hypothetical socioeconomic projection which excludes economic feedbacks of air pollution. This "no-feedback" baseline projection describes hypothetical baseline developments in absence of feedback effects of air pollution on the economy, and is used as the starting point to calculate emissions and concentrations of air pollution, which are then used to assess the impacts and economic feedbacks of the central projection.[1]

The logic of this approach is not to deny that outdoor air pollution is already affecting the economy, but rather to measure the total economic consequences of such air pollution. The no-feedback projection describes the pressures that economic activity puts on the environment, by linking economic activity to emissions and concentrations. The central projection takes the corresponding air pollution impacts, describes how these feed back to the economy and projects the resulting changes in economic activity and specific indicators such as gross domestic product (GDP). The difference in GDP between the two projections reflects the full macroeconomic costs of inaction of outdoor air pollution.

A baseline projection is not a prediction of what will happen, but rather it describes a certain storyline on how key economic and demographic trends affect future economic development in the absence of unexpected shocks. The chosen baseline reflects a continuation of current socio-economic developments, including demographic trends, urbanisation and globalisation trends. The baseline also reflects a continuation of current policies for climate, energy and air pollution (see Box 2.1 for an overview of air pollution policies included in the baseline).

Demographic trends play a key role in determining economic growth. Population projections by age, together with projections of participation and unemployment rates, determine future employment levels. Human capital projections, based on education level projections by cohort, will drive labour productivity. Demographic projections, including effects of changes in fertility, death rates, life expectancy and international migration, are taken from the UN population prospects (2012). The labour force database (participation rates and employment rates by cohort and gender) is extracted from ILO (2011) active population prospects (up to 2020) and OECD Labour Force Statistics and Projections (2011).

Box 2.1. Current air pollution policies included in the baseline

Governments have already implemented a range of policy approaches to limit outdoor air pollution. Information on a large number of economic instruments and voluntary approaches for air pollution can be found in the OECD database on instruments used for environmental policy, at www.oecd.org/env/policies/database. In many countries, so-called "command-and-control" approaches using e.g. regulatory standards are complemented by various economic instruments such as taxes and tradable permit schemes. Voluntary programmes aimed at replacing ovens and heaters, replacing old with LPG and enhanced cook stoves, and retiring old highly-polluting vehicles have also been introduced in recent years in several countries.

In most OECD countries, air pollution policy interventions have become increasingly integrated over the past 10-15 years, helping to increase cost efficiency. Examples include the US Clean Air Act, the Canada-US Air Quality Agreement, the Clean Air Policy Package of the European Commission, and the National Environment Protection Measure for Ambient Air Quality (Australia), all of which have set standards for air quality, focusing on target-setting for a range of air pollutants from stationary sources. These overall frameworks include legislative programmes which target specific sectors, such as power generation, transport, and industrial and residential energy demand. In non-OECD economies, there are fewer examples of cohesive programmes for controlling air pollution. In recent years, much of the focus is on specific policies for controlling emissions from transport, both through standards and economic instruments.

The emission projections presented in this report reflect the effects of current legislations as depicted by International Institute for Applied Systems Analysis (IIASA) in the Greenhouse Gas and Air Pollution Interactions and Synergies (GAINS) model (see Section 2.3 for more details). In principle, all legislation for which information was available is included in the emissions projections, where relevant (e.g. for fuel taxes and congestion charges) through the associated projections of energy use. However, any policy that was not yet fully implemented by late 2012, or that still requires a policy effort to be reached (e.g. the Chinese 11th five-year plan), is excluded from the baseline. This approach provides a snapshot of the effect of policies on current and future emissions; it is a reference point for the assessments of the costs of inaction and the benefits of policy action, and does not reflect a view on the state of very recent and planned environmental policies.

The regional aggregation of ENV-Linkages is used to calculate economic activity, emissions of air pollutants and the feedbacks from pollution impacts on the economy (more detailed representations underlie the calculations of concentrations and biophysical impacts; see Sections 2.4 and 2.5). As shown in Table 2.1, ENV-Linkages distinguishes 12 major countries and 13 groups of countries (regions), based on a mixture of geographical and economic characteristics. For illustrative purposes, some graphs and tables in this report group the underlying 25 regions in 8 "macro-regions", but in all cases the analysis is done at the 25 region level.

Macroeconomic projections for OECD countries are aligned with OECD (2014c). Projections on the structure of the economy, and especially on future sectoral developments, are fundamental for the analysis in this report as they affect the projected emissions of air pollutants. The sectoral assumptions are particularly important as different emission sources are linked to different sectoral economic activities. For instance, final energy demand and power generation affect emissions of a range of pollutants from combustion processes, and in agriculture emissions, especially of NH_3, are linked to the production processes of agricultural goods.

Table 2.1. **Regional aggregation of ENV-Linkages**

Macro regions	ENV-Linkages countries and regions
OECD America	Canada Chile Mexico United States
OECD Europe	EU large 4 (France, Germany, Italy, United Kingdom) Other OECD EU (other OECD EU countries) Other OECD (Iceland, Norway, Switzerland, Turkey, Israel)
OECD Pacific	Oceania (Australia, New Zealand) Japan Korea
Rest of Europe and Asia	China Non-OECD EU (non-OECD EU countries) Russia Caspian region Other Europe (non-OECD, non-EU European countries)
Latin America	Brazil Other Lat.Am. (other Latin-American countries)
Middle East & North Africa	Middle-East North Africa
South and South-East Asia	India Indonesia ASEAN9 (other ASEAN countries) Other Asia (other developing Asian countries)
Sub-Saharan Africa	South Africa Other Africa (other African countries)

Projections of sectoral energy intensities until 2035 are in line with the IEA's World Energy Outlook "Current Policy Scenario" (CPS) (IEA, 2013). After 2035, the IEA trends are extrapolated to fit the macroeconomic baseline thereafter. In fast-growing economies such as the People's Republic of China (henceforth "China"), India and Indonesia, the IEA projects coal use to increase in the coming decades. In OECD regions, however, there will be a switch towards gas, not least in the USA, and this especially in the power generation sector. Further, in OECD economies, energy efficiency improvements are strong enough to imply a relative decoupling of energy use and economic growth, while for emerging economies the decoupling will only be effective in the coming decades. The increase in final energy demand is driven by electricity and by transport; in particular in emerging economies. In line with the trends of the IEA's CPS scenario, electrification of transport modes is assumed to be limited globally.

The projections on agricultural yield developments (physical production of crops per hectare) as well as main changes in demands for crops as represented in the ENV-Linkages baseline are derived from dedicated runs with the International Food Policy Research Institute (IFPRI)'s IMPACT model (Rosegrant et al., 2012) using the socioeconomic baseline projections from ENV-Linkages and excluding feedbacks from climate change on agricultural yields. The underlying crop model used for the IMPACT model's projections is the DSSAT model (Jones et al., 2003). As IMPACT only provides projections to 2050, the trends are linearly extrapolated to 2060. The detailed projections of agricultural production and consumption from IMPACT are then summarised and integrated in ENV-Linkages. According to the projections, while population will increase by 50% from 2010 to 2060, average per capita income is projected to more than double in the same time span. Agricultural production as measured in real value added generated in the agricultural sectors will also more than double by 2060, partially reflecting a shift in diets towards higher-value commodities. The large increase in agricultural production is characterised by a growing share of production in African countries. On the contrary, the market share of OECD countries is projected to decrease.

In principle, feedbacks from climate change on agricultural yields could threaten projected improvements in global food security. Such feedback effects are described extensively in OECD (2015), but excluded from the calculations in this report to allow full focus on the impacts of air pollution. An integrated analysis of both climate and pollution feedbacks is left for future research, but interactions between both themes are discussed in Section 4.2.

2.3. From economic activities to air pollutant emissions

Emissions of air pollutants have been included in the ENV-Linkages model linking them to production activities in different key sectors. The main emission sources are power generation and industrial energy use, due to the combustion of fossil fuels; agricultural production, due to the use of fertilisers; transport, especially due to fossil fuel use in road transport, and emissions from the residential and commercial sectors.

In this study, estimates for selected air pollutants were included: sulphur dioxide (SO_2), nitrogen oxides (NO_x), black carbon (BC), organic carbon (OC), carbon monoxide (CO), volatile organic compounds (VOCs) and ammonia (NH_3). Even if this list does not cover all air pollutants, it includes the main precursors of PM and ground level ozone, which are the main causes of impact on health and on crop yields.

The data on air pollutants used for this report is the output of the GAINS (Greenhouse Gas and Air Pollution Interactions and Synergies) model (Amann et al., 2011 and 2013; Wagner et al., 2007 and 2010; Wagner and Amann, 2009), developed at International Institute for Applied Systems Analysis (IIASA). The GAINS model estimates historic emissions of air pollutants using data from international energy and industrial statistics (not least the EDGAR database), emission factors originating from peer reviewed literature and measurement campaigns, and information about implementation of environmental legislation. Although global coverage and international comparability are most important, the results are compared with the national and international emission inventories that are either published in peer reviewed literature or supplied by countries to the international organisations within existing commitments, e.g. Convention on Long-range Transboundary Air Pollution (LRTAP), United Nations Framework Convention on Climate Change (UNFCCC) Protocol, and EU legislations. The GAINS model structure includes all key known emission sources distinguishing up to about 2000 sector-fuel-technology combinations for each of the 170 countries and regions covered in the model.

The emission projections of the GAINS model used for this project are those relative to the "Current Legislations" (CLE) scenario, which reflects the state of committed air pollution legislation assuming that the required standards can be achieved by existing technologies. These projections are based on activity levels and energy use that reflect those of the 2011 World Energy Outlook (IEA, 2011), but have been rescaled to the more recent energy demand projections of the ENV-Linkages baseline. The projections of the GAINS model used for this project are those that have been prepared for the EU FP7 LIMITS project (see e.g. Rao et al., 2016; Kriegler et al., 2013). The LIMITS project was a large model inter-comparison exercise on interactions between climate policies and other environmental issues, such as air pollution and energy security.[2]

The CLE scenario used in this analysis represents the status of air pollution policies by the end of 2010. Hence, some important developments of the past few years are not captured. The most prominent example is the 11th five year plan in China and the associated legislation; the targets were published already in 2010 but the specific laws and emission limits (more stringent SO_2 and NO_x legislation for the power sector and also for industrial boilers) that are needed for the multi-sectoral assessment of emission factors were introduced later and these could not be considered in the current version of the GAINS scenario.

Emission coefficients have been calculated using the GAINS model projections until 2050. The coefficients are sector- and region-specific to reflect the different implementation rates of respective technologies required to comply with the existing emission legislation in each sector and region. They also change over time to reflect technological improvements, the change in the age structure of the capital stock (more recent generations of equipment submitted to environmental policies replacing the older ones), and the influence of existing policies. Between 2050 and 2060, the emission coefficients (but not total emissions) are assumed to be constant.

The emission coefficients are linked to the projected activity levels to obtain emission projections that are coherent with the economic baseline. Coefficients related to emissions from combustion processes in industrial sectors, transport and residential and commercial energy demand are calculated and linked to the inputs of fossil fuels.[3] Other emissions are linked directly to output (e.g. agricultural goods, cement, metals or waste). Finally, some sources of emissions have been included exogenously in the model as it was not possible to link them to specific economic activities. These are for instance emissions from biofuels. Emissions from forest, agricultural and savannah burning could not be included as they cannot be easily projected to future years. Emissions from aviation and marine bunkers have not been included as they were not part of the GAINS database, although in some coastal regions the effects of marine bunkers on local concentration levels may be significant. This means that, while the main sources of emissions are considered, total emissions of air pollutants have likely been underestimated.

2.4. From emissions to concentrations of air pollutants

Emission projections of precursor gases are used to calculate the associated concentrations of $PM_{2.5}$ and ground level ozone (O_3). High concentrations of $PM_{2.5}$ and O_3 are the drivers of strong impacts on human health and the environment. As discussed in Section 1.3, health impacts caused by NO_2 could not be included in the analysis.

The concentrations of ozone and $PM_{2.5}$ have been calculated using the European Commission Joint Research Centre (EC-JRC)'s TM5-FASST (Fast Scenario Screening Tool) model, which has also been used in e.g. UNEP (2011), in the EU FP7 LIMITS project (Rao et al., 2016; Kriegler et al., 2013) and for the Global Burden of Disease studies (Forouzanfar et al., 2015, and Brauer et al., 2016). TM5-FASST is a reduced form version of TM5 CTM (Krol et al., 2005; Huijnen et al., 2010), a global nested 3-dimensional atmospheric-chemistry-transport model, which simulates ozone and aerosol components with a spatial resolution of 1°×1°.[4] TM5-FASST is based on a set of pre-calculated linear emission-concentration response functions for 56 emitting source regions (Leitao et al., 2015), linking the emissions of precursors SO_2, NO_x, CO, BC, OC, VOCs and NH_3 to the resulting concentrations of pollutants O_3 and $PM_{2.5}$. For further information on TM5-FASST, see Annex B.

While the concentrations are calculated using the ENV-Linkages emission projections as an input, TM5-FASST also includes a fixed natural component from wind-blown dust and sea salt, hence considering both natural and anthropogenic pollution sources. While dust and sea salt are particularly strong in areas with low or no population, they can be carried by winds so it is still important to take them into consideration. Furthermore, TM5-FASST also considers climatic projections in calculating the concentrations, as climatic conditions influence the chemical reactions between pollutants and hence the levels of concentrations. For this project, the RCP8.5 (Riahi et al., 2007) scenario is used. This scenario is the closest to the ENV-Linkages projection of greenhouse gas emissions and average temperature increase and it was previously used as a reference climate scenario for the analysis of the economic consequences of climate change (OECD, 2015).

As impacts are related to exposure, the concentrations are calculated as population-weighted mean concentrations, rather than average concentrations across areas with widely varying population densities. The calculation of the national means of population-weighted $PM_{2.5}$ concentrations is based on combining the spatial concentrations with population maps that approximately reproduce urban background (Rao et al., 2012). The TM5-FASST model also takes into consideration population projections and urbanisation. This is fundamental as the population-weighted concentrations also need to reflect the higher levels of exposure caused by urbanisation.

The TM5-FASST model takes as input the emission projections of the ENV-Linkages model for each of the precursors, regions and sectors considered in the model. The sectoral contributions for each primary pollutant are detailed as much as possible, distinguishing for example between emissions from transport, energy supply and demand, residential and commercial sectors, agriculture, industry and chemicals. This sectoral categorisation is used in the atmospheric model to associate the emissions to specific locations and to estimate the local urban increment from primary $PM_{2.5}$ emissions associated with transport and the residential sector.

A remapping process is used to translate the emission projections for the 25 aggregate regions of ENV-Linkages to the more detailed 56 source regions required for the TM5-FASST model. This is done using available information on emissions from individual countries from a reference gridded emission dataset, in this case RCP8.5 (Riahi et al., 2007), as a proxy for the baseline projection developed in the current study. In a first step, the relative contributions of all countries that are part of a given ENV-Linkages region to the emissions in the RCP8.5's region are used to break down the emissions from the ENV-Linkages' regions to individual countries. In a second step the countries' emissions are re-aggregated to the 56 TM5-FASST source regions.

Concentrations of $PM_{2.5}$ that are used for the calculations of the health impacts are quantified as population-weighted $PM_{2.5}$ values per country. TM5-FASST provides different metrics for ozone impacts. For the O_3 impact on human health, the maximal 6-months mean of daily maximal hourly ozone (M6M) is most appropriate. For damages to crops, an average is taken of the impacts as calculated using AOT40, which is the accumulated hourly ozone above 40 parts per billion (ppb) during a 3-monthly growing season; and using M12, which is the daytime (12 hours) mean ozone concentration during a 3-monthly growing season. These indicators for concentrations of $PM_{2.5}$ and ozone are the starting points to calculate impacts on health and on crop yields.

2.5. From concentrations to impacts on health and agriculture

The following *health impacts* of $PM_{2.5}$ and O_3 were assessed in this analysis: mortality, hospital admissions related to respiratory and cardiovascular diseases, cases of chronic bronchitis in adults and in children ($PM_{2.5}$ only), lost working days ($PM_{2.5}$ only), restricted activity days, and minor restricted activity days due to asthma symptoms ($PM_{2.5}$ only). This selection of impacts is based on the recommendations of the World Health Organization (WHO) under the "Health risks of air pollution in Europe" (HRAPIE) study (WHO, 2013). While this covers a large part of the recognised economic impacts of air pollution on health, there are other impacts that could not be calculated as there is not enough information available (see Chapter 1 for a discussion of other impacts).

The effects of air pollution on health are assessed with concentration-response functions, which link health impacts to the population-weighted mean concentrations of $PM_{2.5}$ and O_3. Concentration-response functions are typically estimated by gathering data on the occurrence of the health impacts, and running regressions that relate them to population-weighted concentrations of air pollutants, controlling for factors such as temperature, relative humidity, wind speed or season.

To obtain projections of the impacts of air pollution on health, it is also necessary to understand future levels of exposure. Information is needed on population projections, as well as the demographic structure of the population and its expected development over time. The calculation of health impacts has been done based on UN's demographic and population projections (2012), in line with the data used for the ENV-Linkages baseline and the OECD's long-term macroeconomic projections (OECD, 2014c).

For the base year, 2010, the impacts of $PM_{2.5}$ on mortality assessed in this study are based on the results of Forouzanfar et al. (2015) and Brauer et al. (2016).[5] Effects of ozone on mortality in 2010 are based on the earlier results of Lim et al. (2012) and Burnett et al. (2014). While updated results for the health impacts of ozone are available in Forouzanfar et al. (2015), the results in this report are based Lim et al. (2012). Given the dominance of the impacts of PM2.5 using older estimates for ozone only marginally affects the total results on the total costs of outdoor air pollution calculated in this report.

Forouzanfar et al. (2015) adopt a non-linear response function for PM mortality, with the rate of increase of mortality declining as PM concentrations rise (see Box 2.2 for an overview of these Global Burden of Disease studies). This assumption has been followed to generate lower projections of mortality. Upper projections are based on a linear relationship between mortality and concentrations. The use of a range recognises potentially significant uncertainty in the development of the non-linear relationship.

Quantification of morbidity effects requires different data, including the concentration-response relationship, the size of the population at risk, and the prevalence of morbidity. As this level of information was available for only a small number of countries, the quantification of morbidity effects is based on extrapolation of the results of studies performed for the Clean Air Policy Package of the European Commission (Holland, 2014a; European Commission, 2013) where the HRAPIE recommendations of WHO (2013b) were implemented as multipliers on the all-cause mortality from pollutant exposure. To ensure consistency, a correction was applied to account for differences between quantified all-cause deaths from Holland (2014a) and cause-specific mortality estimates from Forouzanfar et al. (2015). Ideally changes in behaviour (e.g. in diet, smoking habits, etc.), social changes (e.g. healthcare and employment) and medical changes (e.g. changes in healthcare systems and in treatment of diseases) over time and between world regions, should be factored into the analysis, but this is not possible owing to lack of data at global level. Further details on the methodology used to calculate health impacts are presented in Annex C.

Crop yield changes have been estimated following the methodology described in Van Dingenen et al. (2009). Crop losses for rice, wheat, maize and soybean are calculated in TM5-FASST based on concentrations of ozone during the growing season.[6] Gridded growing season and crop yield data are obtained from the Global Agro-Ecological Zones (GAEZ, version 3) (FAO/IIASA, 2012). For wheat and rice, growing season data are available for different varieties (spring wheat, winter wheat/dryland rice, wetland rice); however, yield data are provided for total wheat only. For maize and soybean, only one growing season dataset is available. Yield losses have been calculated assuming either that all wheat is spring wheat or that it is all winter wheat. The same assumptions have been taken for rice. The calculations in this report have been made with average values between the two assumptions on crops being all spring or all winter; a sensitivity analysis is presented in Chapter 4. It should be acknowledged that the projected crop yield changes are less robust than the projections of health impacts, owing to a much smaller underlying scientific literature. To ensure consistency with the crop yield projections in ENV-Linkages, the crop yield changes are expressed as a percentage change from the ENV-Linkages no-feedback projections.

Box 2.2. **The Global Burden of Disease studies**

The Global Burden of Diseases, Injuries, and Risk Factors Study (GBD) provides a methodology to quantify health loss from hundreds of diseases, injuries, and risk factors. GBD is the largest and most comprehensive effort to date to measure epidemiological levels and trends worldwide (www.healthdata.org/gbd).

The GBD initiative dates back to the early 1990s, when the World Bank commissioned the original GBD study (World Development Report 1993: Investing in Health). GBD work was institutionalised at the World Health Organization (WHO), and the organisation continued to update GBD findings.

The next comprehensive GBD update, the Global Burden of Diseases, Injuries, and Risk Factors Study 2010 (GBD 2010) published new estimates for the complete time series from 1990 to 2010 and an explanation of its methods in The Lancet in December 2012 (see Lim et al., 2012). While earlier work had been conducted mainly by researchers at Harvard and the WHO, GBD 2010 brought together a community of nearly 500 experts from around the world in epidemiology, statistics, and other disciplines.

With the Institute for Health Metrics and Evaluation (IHME) as the co-ordinating centre for an international network of GBD contributors, the entire time series of GBD estimates is being updated regularly to provide detailed information on population health (IHME, 2015). The first update, GBD 2013 (see e.g. Forouzanfar et al., 2015 and Brauer et al., 2016), expands the methodology, datasets, and tools used in GBD 2010 and presents estimates of all-cause mortality, deaths by cause, years of life lost, years lived with disability, and disability-adjusted life years by country, age and sex. GBD 2013 produced estimates for 323 diseases and injuries, 67 risk factors, and 1 500 sequelae for 188 countries. It reflects the work of more than 1 000 researchers in more than 100 countries.

Crop yield changes for those crops that are not covered by the calculations with TM5-FASST are projected using the information in Mills et al. (2007), following the methodology of e.g. Chuwah et al. (2015): yield changes for these crops are based on their relative sensitivity to ozone as compared to rice. For instance, Mills et al. find that sugar is roughly 1.5 times as sensitive as rice, and thus for each region the projected yield impacts in ENV-Linkages are also assumed to be 1.5 times those of rice. While necessarily very crude, this approach ensures that all crops are covered and avoids major distortions in the projections that might result from missing data.

2.6. Unit values for the analysis of health impacts

The valuation of the health impacts of outdoor air pollution includes both mortality and morbidity. Total health costs can be calculated by multiplying the impacts for each endpoint considered (e.g. number of hospital admissions, cases of illness, and premature deaths) by appropriate estimates of the unit value of each impact (e.g. the economic value of a hospital admission, a case of illness, and a premature death).

Different techniques are available to establish unit values. They can be estimated through a cost-of-illness approach and/or through direct monetary valuation techniques such as stated preference (SP) or revealed preference (RP) methods to assess the willingness-to-pay (WTP) to reduce environmental risks. Cost of illness and direct valuation techniques are often used in different contexts. The cost-of-illness approach is generally used in cost-effectiveness analysis (CEA) in order to provide an economic rationale for the rationing of health care resources in specific policy or programme proposals. In this instance, the benefits of investing in such resources are expressed in terms of the number of cases of illness avoided, or an index such as the number of "quality adjusted life years" (QALYs) gained. In contrast, WTP measures of these benefits are often used by economists in cost-benefit analysis (CBA), where the total costs and benefits of projects and policy proposals can be compared using a common money metric.

The cost-of-illness approach estimates the societal burden of disease by quantifying all costs related to illness that can be linked to market or financial transactions. These include "direct costs" (e.g. healthcare costs, expenditures in medicines and medical supplies) and "indirect costs" (e.g. the value of lost productivity because of reduced working time). The cost-of-illness approach does not take into consideration any of the costs that do not have a marketable or tradable value, such as the costs of pain and suffering. Using this approach disregards a potentially significant part of the loss to people related to mortality and morbidity. For example, using a cost-of-illness approach, a premature death would be evaluated with the future production potential of the deceased person, hence ignoring other aspects of premature death and the associated monetary values.

Stated and revealed preference techniques, by contrast, usually aim at estimating the welfare costs of illness or risk of premature death, often focusing on the non-market costs. SP methods (such as contingent valuation or choice modelling) rely on surveys to ask respondents for their WTP to reduce their mortality risk. RP methods use market behaviour to reveal individual preferences. In particular "hedonic pricing" methods are based on individuals' behaviour in markets where prices reflect differences in mortality risk (e.g. a labour market, where wages reflect differences in workplace mortality risks), and "averting costs" methods are based on markets for products that reduce mortality risks (e.g. buying motorcycle helmets to reduce mortality risks in traffic accidents).

Both SP and RP methods have their strengths and weaknesses, but there has been a growing emphasis on SP methods in recent years (OECD, 2012), especially in the context of environmental impacts. While these techniques are very useful in the evaluation of total economic costs of health or environmental impacts, they are generally not as accurate as cost-of-illness estimates. For example, stated preferences techniques – based on the responses to surveys – potentially introduce a number of biases and difficulties. One of the main difficulties to consider when using results from SP surveys is that respondents to surveys on the willingness to pay for a reduction in the risk of dying prematurely, may have different background or initial risks (i.e. the perceived risk of "dying anyway"). Providing informing factors can limit this issue. Perhaps the main potential bias is that responding to a survey does not involve a real commitment to pay what is stated in the survey – it is hypothetical only. For a comprehensive overview of the characteristics and shortcomings of the valuation literature see OECD (2006) and OECD (2012).

While the cost-of-illness and direct valuation approaches both aim to associate an economic cost to episodes of illness, the estimates of the two methodologies largely differ, as they measure two different aspects of the same issue. An example of the difference in measurement is provided by Chestnut et al. (2006), who estimate the economic benefits of reducing respiratory and cardiovascular hospitalisations based on both cost-of-illness and SP. The WTP estimates indicate that individuals value prevention of a five-day hospitalisation event at an average of approximately USD 2 400, while the average total cost-of-illness estimates per hospitalisation are USD 22 000-39 000.

Combining the two methodologies poses challenges in terms of double counting and comparability of estimates, but it can also help better assess the full societal costs of air pollution. Stieb et al. (2002), combine empirical data on the duration and severity of episodes of cardiorespiratory disease with cost-of-treatment, lost productivity, and WTP to avoid acute cardiorespiratory morbidity outcomes linked to air pollution.

Willingness-to-accept (WTA) is an alternative technique to WTP to attribute monetary values to mortality and to the disutility of illness. Using WTA generally provides larger estimates (Horowitz and McConnell, 2002), in part because the respondents to a WTA survey are not bounded by income. This means that, especially in the context of mortality risk valuation, respondents to surveys could provide unrealistically large values. Further, there is often a large share of "don't know" and protest responses when respondents are asked to accept an increase in mortality risk (OECD, 2012). OECD (2006) presents a detailed comparison of both concepts, and provides theoretical and practical reasons for using WTP.

Establishing unit values for mortality

The valuation of mortality impacts in this report relies solely on results from SP studies. In particular, it is based on estimates of the "value of a statistical life" (VSL) (see Box 2.3 for a discussion on valuing premature deaths due to air pollution). This is a long-established metric, which can be quantified by aggregating individuals' WTP to secure a marginal reduction in the risk of premature death over a given timespan (see OECD, 2012 and OECD, 2014a). Using solely direct monetary valuation means that certain indirect costs related to premature deaths are possibly not considered. Respondents to surveys on the risks of dying prematurely are unlikely to consider costs such as those related to the economic repercussions of lost productivity on the economy (for the working population). These are, however, likely to be a minor component of the value that can be associated to the premature death of an individual.

Box 2.3. **Valuing premature deaths with the value of a statistical life**

One of the most common procedures to value risks to life in standard economic theory is the value of a statistical life (VSL) (OECD, 2006). The VSL is derived from aggregating individuals' WTP to secure a marginal reduction in the risk of premature death over a given timespan.

The VSL is most commonly elicited through stated preference techniques, although revealed preferences techniques are also used. Alberini et al. (2016) provides an overview of the different methodologies used to elicit the VSL as well as their characteristics and shortcomings.

Box 2.3. **Valuing premature deaths with the value of a statistical life** *(continued)*

OECD (2012) describes the basic process for deriving a VSL from a state preference survey. Suppose the survey finds an average WTP of USD 30 for the reduction in annual risk of dying from air pollution from 3 in 100 000 to 2 in 100 000. This means that each individual is willing to pay USD 30 to have this 1 in 100 000 reduction in risk. In this example, for every 100 000 people, one death would be prevented with this risk reduction. Summing the individual WTP values of USD 30 over 100 000 people gives the VSL – USD 3 million in this case.

It is important to emphasise that the VSL is not the value of an identified person's life, but rather an aggregation of individual values for small changes in risk of death (OECD, 2012). As such, the total economic cost of the impact equals the VSL multiplied by the number of premature deaths; the economic benefit of a mitigating action becomes the same VSL multiplied by the number of lives saved (OECD, 2014a).

One large debate in the use of VSL is how the age of individuals matters in relation to different risk contexts. The same VSL is easily applicable in contexts in which the risk of premature deaths is reduced to the same extent for populations of all ages. In cost-benefit analysis exercises for policies that specifically focus on children's health, it is preferable to use specific values to evaluate the policy benefits for children (OECD, 2010). There are, however, difficulties in establishing child-specific VSL values since it is not possible to use surveys to elicit children's own preferences and biases, such as altruism, may arise when adults are asked to value risks for their children. In cases of evaluation of regulations targeted to reducing children's health risks, OECD (2012) and Lindhjem and Navrud (2008) suggest that VSL for children should be a factor of 1.5-2.0 higher than adult VSL. Air pollution is found to lead to premature deaths mostly of elderly people and, to a smaller extent, of children (WHO, 2014). Nevertheless, mortality risks, which are the ones considered in this report, mostly affect the elderly and the contribution from acute respiratory deaths in children (younger than 5 years of age) is very small. An adjustment is therefore not needed in the calculations of this report.

Age can also be taken into consideration by using the "value of a life year lost" (VOLYs), sometimes described as "value of a statistical life year" (VSLY). This technique calculates the number of "years of life lost" (YOLLs) owing to a specific risk and based on an estimated life expectancy, and then evaluates them by multiplying them by the VOLY. One issue with this technique is that the combination of counting YOLLs, rather than lives lost, means that the VOLY approach "explicitly places a lower value on reductions in mortality risk accruing to older populations with lower quality of life" (Hubbel, 2002). While there is a general agreement that children's health risks should ideally be valued differently, there is little support for the differentiation for adults of different ages. Further, VOLYs are rarely derived from surveys (Hunt, 2011). There are also major complications in the robust estimation of YOLLs, and the extent to which existing country-specific life expectancy values can and should be used. YOLLs can be calculated using country-specific life tables which are provided by the UN World Population Prospects (UN, 2015), although this requires elaborated calculations to obtain YOLLs for all world regions. The Global Burden of Disease studies define YOLLs as the difference between the age at death minus the global "longest possible life expectancy" (www.healthdata.org/gbd/faq) when calculating the numbers of years gained by avoiding a premature death. Using this assumption implies that especially in countries that currently have relatively low life expectancy, the total number of YOLLs will be greatly overestimated. In such cases it is possible that the valuation of premature deaths through a large number of VOLYs become significantly larger than when using VSL. Nevertheless, costs of premature deaths are usually higher when calculated using VSL. Given the limitations of the use of VOLYs, and following OECD (2012, 2014a), in this report the premature deaths are evaluated with the same VSL for all age groups.

Box 2.3. **Valuing premature deaths with the value of a statistical life** *(continued)*

One further issue when using VSL in the context of health impacts caused by air pollution is latency, namely the difference between time of exposure and the actual impact (premature death). The effect of latency on WTP is theoretically undetermined (OECD, 2012). Economic theory is usually based on the principle that people discount the future at a positive rate. Their utility will also vary with different periods of life in a way that can make WTP to reduce future mortality risks higher than their WTP to reduce immediate risks (see e.g. Hammitt and Liu, 2004). The meta-analysis in OECD (2012) was used to study whether VSL estimates systematically vary with different characteristics of the valuation methodology employed, characteristics of the change in mortality risk (e.g. type of risk, latency, cancer risk etc.), socio-economic characteristics of the respondents and other variables. Based on the literature review and the meta-analysis, OECD (2012) concludes that no adjustments should be made for latency in base VSL values.

OECD (2014a) provides country-specific VSL values for adults for OECD Member countries and some non-OECD economies, while OECD (2014b) does so for countries in the South and South East Asia region. As this report has global coverage, it was necessary to calculate VSL values for countries not covered by previous OECD studies. This was done using the benefit transfer methodology based on average national income, as outlined in OECD (2012) and detailed in Box 2.4. The key parameter in this methodology is the elasticity of income, which determines the extent to which the VSL changes according to different income levels. In this report, the income elasticity used for the calculations if 0.8 high income countries, 0.9 for middle-income countries and 1 for low-income countries. To analyse the sensitivity of the results to the chosen values of the income elasticity, alternative elasticity values are considered (see Section 5.1).

Box 2.4. **Benefit transfer for the value of a statistical life**

OECD (2012) provides a methodology to calculate country-specific VSL based on average national income through a benefit transfer methodology. In units of 2005 USD, the indicated range for OECD countries is USD 1.5-4.5 million, and the recommended base value is USD 3 million. Reference VSL values for OECD in 2005 are obtained from a rigorous meta-analysis of VSL studies (OECD, 2012). Starting with 1 095 values from 92 published studies, OECD-recommended VSL values were calculated for an average adult.

As argued in OECD (2006 and 2014a), WTP varies with income and income is one of the main indicators used in preference-based technique for measuring VSL. Country-specific VSL values are calculated starting from a reliable estimate for a specific region, in this case the OECD base value of USD 3 million, and then adjusting the VSL for other countries based on income levels. The use of a local VSL reflects the situation that the valuation is done in a specific country; this is appropriate as both the costs and benefits of air pollution and pollution control policies are largely within the same region (OECD, 2014a). This is in contrast to e.g. climate change, where the use of a different VSL for mortality in different countries is very controversial, as the beneficiaries of a policy are largely located in other countries (as greenhouse gases are uniformly mixing in the atmosphere).

Box 2.4. **Benefit transfer for the value of a statistical life** *(continued)*

Several studies attempt to evaluate the income elasticity of the WTP to reduce the risk of premature death. The meta-analysis in OECD (2012) finds that the income elasticity is in the range of 0.7-0.9 for OECD countries, with significantly higher income elasticities for countries in the bottom 40th percentile of income. Longitudinal studies provide additional evidence that WTP varies at different stages of economic development (Hammitt and Robinson, 2011). In particular, the range proposed in OECD (2012) has been judged to be too low for low income countries as using such values would imply unrealistically high WTP values for these countries. Given this evidence, this report uses an elasticity of 0.8 for high-income countries, 0.9 for middle-income countries and 1 for low-income countries (country groups are distinguished using the World Bank income thresholds).

This benefit transfer methodology is used to adapt VSL to other countries, but also to estimate its growth over time. As argued in OECD (2006), income should be used as the reference variables also to adapt WTP over time, so as to avoid situations in which for instance the WTP to save a statistical life rises faster over time than the rate of inflation. Existing studies, such as Costa and Kahn (2004) who calculate the VSL changes in the US for the period 1940-80, find that VSL rises over time as income rises.

Country- and year-specific VSL is calculated following this formula:

$$VSL_r^t = VSL_{OECD}^{2010} \cdot \left(\frac{Y_r^t}{Y_{OECD}^t}\right)^{\beta}$$

where:

Y is the average income (GDP per capita) of country r in year t expressed in 2010 USD PPP;

β is the income elasticity of VSL. It measures the percentage increase in VSL for a percentage increase in income.

This methodology is applied in this analysis to obtain VSL values for the all countries in the world, as well as for the projections to 2060. The extrapolations are based on the projected country-specific income values. The income projections used are the same as those used to calibrate the ENV-Linkages model: IMF Economic Outlook (2014) until 2017 and then on the economic projections of the ENV-Growth model (Dellink et al., 2016).

Establishing unit values for morbidity

The valuation of morbidity in this report combines a separate evaluation of cost-of-illness (healthcare and labour productivity costs) and welfare costs.[7] The literature on the costs associated with air pollution effects on the demand for healthcare is very sparse compared with that which seeks to provide estimates of the overall economic cost of air pollution on health. Discussion in the literature often misrepresents estimates of total cost as being the costs of healthcare or of healthcare and productivity losses, when these are only a few components of the total costs of outdoor air pollution. For the purpose of this work, the term "healthcare costs" is specific to the costs incurred in treating illnesses, while the costs of discomfort, pain and suffering related to illness are referred to as "disutility costs" or "welfare costs".

As already discussed, healthcare costs can be evaluated using the cost-of-illness approach. While data availability is certainly an issue, a quantification of healthcare costs is at least theoretically straightforward, as they are linked to market transactions and thus have established, observable prices.[8] Nevertheless it is not easy to establish a reference unit value for healthcare costs, as they vary substantially across different countries, owing to the differences in healthcare systems, but also in the way people face illness. Even within the same continent there can be large variations. For instance, healthcare costs for chronic bronchitis have been estimated in a series of European studies using similar methods to be EUR 530/patient/year in France (Piperno et al., 2003) but EUR 3 238/patient/year in Spain (Izquerdo, 2003). Unit values for healthcare expenditures have been established for the OECD based on Holland (2014a). Country-specific unit values are then calculated based on the relationship between healthcare expenditure and GDP per capita, using the World Bank's 2015 total healthcare expenditure as a percentage of GDP at the national level (World Bank, 2015).

Welfare costs, which include the disutility costs of illness related for example to pain and suffering, are evaluated using available results on WTP from SP studies. In particular, this report uses the values calculated for the European Commission (Holland, 2014a) as a starting point to establish unit values for the welfare costs of the morbidity endpoints. Extension of morbidity welfare costs to specific countries uses benefit transfer based on income, as for mortality costs (see Box 2.4). There is a potential bias in transferring estimates of the disutility of morbidity from existing studies, mostly developed in Europe, to the global context. Preferences on health and the valuation of illness can greatly vary between different countries. For example, Ready et al. (2004) illustrate that using international transfer of unit values in the evaluation of the benefits of specific health impacts introduced a transfer error even between European countries. In the context of a global study, however, benefit transfer is the only available technique, as the availability of valuation studies on the impacts of air pollution are only focused on a few areas of the world.

Resulting unit values

The unit values used are presented in Table 2.2 for each health endpoint, including a breakdown to the different cost elements (welfare and healthcare costs).[9] The value used for mortality is USD 3 million, following OECD (2014a). The morbidity values are established based on (Holland, 2014a).

Table 2.2. **Unit values used in the analysis of the health impacts**

USD, 2005 PPP exchange rates

Effect	Cost element	Value
Mortality, premature deaths	Welfare cost	3 million
Chronic bronchitis in adults (new cases)	Welfare cost	61 610
	Healthcare cost	13 070
Bronchitis in children (cases)	Welfare cost	680
	Healthcare cost	57
Equivalent hospital admissions (respiratory and cardiovascular diseases)	Welfare cost	575
	Healthcare cost	3 430
Restricted activity days	Welfare cost	106
Minor restricted activity days (asthma symptom days)	Welfare cost	48

Note: Values are for the OECD. They are unit values and as such they refer to costs per statistical life, case of illness, hospital admission and day with restricted activity.

Source: Own evaluation based on Holland (2014a).

The methods adopted leave little potential for double counting of the different elements of valuation of morbidity as costs are fully attributed to the main cost component. Mortality is only associated with welfare costs and it is not included in the modelling analysis of market costs. Unit values for chronic bronchitis in adults, bronchitis in children and hospital admissions are established for both welfare and healthcare costs. While respondents to surveys that are used to derive welfare costs may also consider some of the market costs in their answers, the largest share of the costs are likely to be related to non-market costs. The SP studies used to establish the unit values for the disutility of illness have been conducted in countries with well-functioning public health care systems, which reduce the risk of respondents including components other than disutility when they state their WTP for avoiding clearly specified episodes of illness. The values used should therefore reflect different and complementary aspects of illness.

For lost working days, the assumption is that the main impact is reduced productivity, while for (minor) restricted activity days discomfort is assumed to dominate. Hence, lost working days impact labour productivity and are included in the calculations of market impacts with the ENV-Linkages model. Costs associated with (minor) restricted activity days on the contrary are assessed through their welfare costs. Annex C further discusses double counting issues.

Once the unit values are established, the overall healthcare costs, welfare costs of illness and of mortality can be calculated by multiplying the number of cases of illness and of premature deaths by the unit values (see Section 2.8). The overall costs are therefore an aggregate of average individual costs for the affected individuals. The overall healthcare costs are used as an input to the ENV-Linkages model to calculate the market costs of outdoor air pollution. The market costs then include direct costs related to the overall health expenditures as well as indirect costs originating from the repercussions on consumption, savings, production and other economic activities (see Section 2.7).

2.7. From impacts to consequences for economic growth

The market impacts are modelled directly in ENV-Linkages following a production function approach. This means that market impacts are not assumed to only affect macroeconomic variables such as GDP, but to directly affect specific elements in the economic system, such as labour productivity or land productivity. The impacts are thus modelled as changes in the most relevant parameters of the production function underlying the model structure. The resulting changes in the economy (both at sectoral and macroeconomic level) are expressed as percentage change with respect to the projection without feedbacks to the economy (cf. Section 2.2). They are calculated for each time period up to the time horizon (2060) and thus reflect the annual economic consequences that result from the stream of impacts over time. The scenario which includes the market impacts from air pollution is referred to as the *central projection.*

Three market impacts are included in the model: changes in health expenditures due to increased incidence of illnesses, changes in labour productivity due to increased incidence of illnesses, and changes in agricultural crop yields. Table 2.3 summarises the impacts modelled and the data sources.

Changes in health expenditures are implemented in the model as a change in demand for the aggregate non-commercial services sector. The amount of additional health expenditures introduced in the model is calculated multiplying the number of cases of illnesses and of hospital admissions by the unit values for healthcare specified in Section 2.6. It is assumed that the additional health expenditures affect both households and government expenditures on healthcare.[10] The extent to which households or governments are affected depends on regional characteristics of the health system in terms of their relative contribution to healthcare. The distinction between households and government expenditures has been done using World Bank data on the proportion of healthcare expenditures paid by households and by the government (World Bank, 2015). A close relationship is noted between healthcare expenditure and GDP per capita for all but a few countries (World Bank, 2015), facilitating extrapolation of data on specific health endpoints between countries.

Table 2.3. **Air pollution impacts included in ENV-Linkages**

Impact categories	Impacts modelled	Data sources
Health	Changes in health expenditures due to changes in incidences of bronchitis, respiratory and cardiovascular diseases, etc.	Calculations based on Holland (2014a) and on results from the Global Burden of Disease studies (Forouzanfar et al., 2015, and Brauer et al., 2016 for PM; Lim et al., 2012, and Burnett et al., 2014 for ozone).
	Changes in labour productivity due to lost working days caused by changes in incidences respiratory and cardiovascular diseases.	
Agriculture	Changes in crop yields	Calculations by the EC-JRC Ispra with the TM5-FASST model (Van Dingenen et al., 2009).

Changes in labour productivity are directly implemented in the model as percentage changes in the regional productivity of the labour force. Productivity losses are calculated from lost working days, following the methodology used in Vrontisi et al. (2016), using assumptions on the average number of work days per year in each region (World Bank, 2014). The approach to reduce labour productivity rather than labour supply is more appropriate when the dominant effect of the illness is to reduce average output per worker, rather than total labour costs borne by employers. This holds especially when employees are compensated for sick leave, or when workers show up to work while being ill (presenteeism).

Changes in crop yields are implemented in the model as a combination of changes in the productivity of the land resource in agricultural production, and changes in the total factor productivity of the agricultural sectors. This specification, which is in line with OECD (2015), mimics the idea that agricultural impacts affect not only purely biophysical crop growth rates but also other factors that affect output, such as the effectiveness of other production inputs. Air pollution affects crop yields heterogeneously in different world regions, depending on the concentrations of ground level ozone.

Once impacts on crop yields, health expenditures and labour productivity have been included in ENV-Linkages, the model is used to calculate the macroeconomic costs of air pollution in the central projection. These costs are the result of the direct market costs as well as the adjustment processes that take place in the model (indirect market costs). For instance, an increased demand for healthcare may result in a lower demand for other services, while changes in crop yields for certain crops may result in changes in production of other substitute crops and even other sectoral activities as well as changes in trade patterns.

2.8. From impacts to welfare costs

The last step in the analysis is the assessment of the welfare costs of outdoor air pollution. For non-market costs related to health impacts, these are calculated by multiplying the results related to the relevant health endpoints (step 4) with the appropriate unit values (step 5). More precisely, for mortality, the number of premature deaths is multiplied with the value of a statistical life (VSL). Similarly, for each of the morbidity endpoints, the results are multiplied with the corresponding unit value to calculate the welfare costs related to the disutility of illness.

The analysis of the economic consequences of the market impacts is done with a focus on the most common indicator of economic activity, GDP. This is also the reference point to investigate the consequences of air pollution on economic growth. The market costs are also expressed in terms of welfare to facilitate the comparison with other cost components. This is done using the equivalent variation of income, which is a common measure of welfare impacts of a shock in a general equilibrium framework. It measures the change in income that, at initial prices, would have the same welfare effect as the changes induced by the shock to the system (Hicks, 1939). Thus, the welfare costs of market impacts are represented as a change in income, in constant USD. The equivalent variation represents the maximum willingness to pay to avoid the deterioration in the welfare of consumers (this is known in the economics literature as Hicksian equivalence).

Finally, market and non-market welfare costs can be compared and aggregated to provide an assessment of total welfare costs. Having different methodologies to calculate market and non-market costs complicates the possibilities to aggregate numbers. However, market and non-market costs can be added when both are expressed in terms of aggregate income losses, and using the same metric, i.e. constant 2010 USD using PPP exchange rates.

One further issue comes with aggregating welfare costs across countries and regions. In principle, equity weights can be used to create a social welfare function that affect how a trade-off between welfare changes in different countries is measured. Such weights could be used in establishing VSL and morbidity values for developing countries, and in the welfare measures used in the general equilibrium model. The effect of welfare weights is that they provide a "fairer" measure of the global social welfare associated with the welfare costs presented in this report; they would also reflect that the marginal utility from an additional unit of income is larger in poorer countries than in rich countries. However, this report abstains from introducing welfare weights for two reasons. First, the aim of the report is not to find a socially optimal level of pollution; rather, it aims at highlighting the regional consequences of unmitigated outdoor air pollution. Although the regional results are sometimes aggregated to a global total, that is purely for illustrative purposes. A second reason not to adopt equity weights is that these reflect essentially a moral judgement and it is extremely difficult to find appropriate welfare weights that would be uncontroversial. Finally, equity weighting introduces a new level of complexity in the results that is avoided by focusing on the results expressed in terms of *income* changes.

Notes

1. In principle, the feedback effects will affect emission levels and thus one should iterate between the central projection and the no-damage projection until consistency is reached on the level of emissions. This iterative process is, however, very computationally expensive, and only relevant when the emission levels in the central projection are significantly different from those in the no-damage projection.

2. This dataset has also been used as a basis for the Energy Modelling Forum (EMF) 30 model comparison exercise, whose output has been used to check on the robustness of the implementation of the air pollutants in the ENV-Linkages model.

3. Ideally, for transport, it would be better to consider fuel use per kilometre or passenger, but the ENV-Linkages model does not include such details.

4. The reduced-form version TM5-FASST mimics the full set of chemical, physical and meteorological processes represented in TM5-CTM, for the meteorological year 2001. They represent the formation of secondary ammonium sulphate and nitrate from SO_2, NO_x and NH_3 emissions, the formation of O_3 from NO_x and VOC and the transport and wet and dry removal of all pollutants from the atmosphere.

5. By building on the GBD studies, the implicit weaknesses of those studies are included also here. For instance, there may be a risk that interactions between air pollution and tobacco smoking are not adequately addressed in attributing mortality to outdoor air pollution. Nonetheless, the GBD studies provide the most robust and comprehensive information available for assessing the impacts of air pollution on mortality at a global level.

6. Rice, wheat, maize and soybean represent more than half the total volume of global agricultural production, but less than half of the value.

7. It is also possible to distinguish the morbidity costs of the health impacts of air pollution into *(i)* resource costs, which are represented by the direct medical and non-medical costs associated with treatment for the adverse health impact of air pollution plus expenditures on averting behaviour; *(ii)* opportunity costs, which are associated with the indirect costs related to loss of productivity and/or leisure time due to the health impacts; and *(iii)* disutility costs, which refer to the pain, suffering, discomfort and anxiety linked to the illness. The analysis of this report covers each of these three types of impacts at least partially, as resource costs relate to health expenditures, opportunity costs to labour productivity changes and disutility costs are included in the welfare cost evaluation.

8. For regions where healthcare costs cannot be directly assessed, results for other regions have been extrapolated.

9. For consistency with original sources, the figures in this table are given in 2005 USD. These have then been converted to 2010 USD in the modelling framework. Results from the analysis are also presented in 2010 USD.

10. In reality, private sector business also plays a role in the supply of healthcare through employer-based insurance. These expenditures are not considered separately in the modelling framework. Further, an alternative assumption on governments and households, is that they could decide not to increase their health expenditures and accept a lower level of health care. Such a response will, however, likely result in larger welfare costs. The approach used here can therefore be seen as a lower bound for the health costs.

References

Alberini et al. (2016, forthcoming), "Approaches and Issues in Valuing the Cost of Inaction of Air Pollution on Human Health", *OECD Environment Working Papers*, No. 108, OECD Publishing, Paris.

Amann, M. et al. (2013), "Regional and Global Emissions of Air Pollutants: Recent Trends and Future Scenarios", *Annual Review of Environment and Resources*, Vol. 38, pp. 31-55.

Amann, M. et al. (2011), "Cost-effective control of air quality and greenhouse gases in Europe: Modeling and policy applications". *Environmental Modelling & Software*, Vol. 26(12), pp. 1489-1501.

Brauer, M. et al. (2016), "Ambient Air Pollution Exposure Estimation for the Global Burden of Disease 2013", *Environmental Science and Technology*, Vol. 50, pp. 79-88.

Burnett, R.T. et al. (2014), "An integrated risk function for estimating the global burden of disease attributable to ambient fine particulate matter exposure", *Environmental Health Perspectives*, Vol. 122, pp. 397-403.

Chateau, J., R. Dellink and E. Lanzi (2014), "An Overview of the OECD ENV-Linkages Model: Version 3", *OECD Environment Working Papers*, No. 65, OECD Publishing, Paris, http://dx.doi.org/10.1787/5jz2qck2b2vd-en.

Chestnut et al. (2006), "The Economic Value of Preventing Respiratory and Cardiovascular Hospitalizations", *Contemporary Economic Policy*, Volume 24(1), pp. 127-143.

Chuwah, C. et al. (2015), "Global impacts of surface ozone changes on crop yields and land use", *Atmospheric Environment*, Vol. 106, pp. 11-23.

Dellink, R.B. et al. (2016, forthcoming), "Long-term economic growth projections in the Shared Socioeconomic Pathways", *Global Environmental Change*, http://dx.doi.org/10.1016/j.gloenvcha.2015.06.004.

European Commission (2013), "The Clean Air Policy Package", Commission Staff Working Document. Impact Assessment, No. SWD(2013)531, European Commission, Brussels, http://ec.europa.eu/governance/impact/ia_carried_out/docs/ia_2013/swd_2013_0531_en.pdf (last accessed on 3 May 2016).

IHME (Institute for Health Metrics and Evaluation) (2015), *Global Burden of Disease Compare,* Seattle: WA: IHME.

IIASA/FAO (International Institute for Applied Systems Analysis/Food and Agriculture Organisation) (2012), "Global Agro-ecological Zones (GAEZ v3.0)". IIASA, Laxenburg, Austria and FAO, Rome, Italy, http://webarchive.iiasa.ac.at/Research/LUC/GAEZv3.0/ (last accessed on 3 May 2016).

Forouzanfar, M.H. et al. (2015), "Global, regional, and national comparative risk assessment of 79 behavioural, environmental and occupational, and metabolic risks or clusters of risks in 188 countries, 1990-2013: a systematic analysis for the Global Burden of Disease Study 2013", *The Lancet*, Vol. 386(10010), pp. 2287-2323.

Global Burden of Disease (GDB) (2010), "GBD 2010 country results: a global public good, GBD 2010 Country Collaboration", *The Lancet*, Vol. 381(9871), pp. 965-970.

Hammitt, J.K. and J.-T. Liu (2004), "Effects of Disease Type and Latency on the Value of Mortality Risk", *The Journal of Risk and Uncertainty*, Vol. 28, pp. 73-95.

Hammitt, J.K. and Robinson, L.A. (2011), "The Income Elasticity of the Value per Statistical Life: Transferring Estimates between High and Low Income Populations", *Journal of Benefit-Cost Analysis*, Vol. 2, ss. 1, Article 1, pp. 1-27.

Hicks, J.R. (1939), *Value and capital: An inquiry into some fundamental principles of economic theory*, Clarendon Press, Oxford.

Holland, M. (2014), Cost-benefit analysis of final policy scenarios for the EU Clean Air Package. Corresponding to IIASA TSAP report no.11, Version 2a. Report to European Commission DG Environment. March 2014, http://ec.europa.eu/environment/air/pdf/TSAP%20CBA.pdf (last accessed on 3 May 2016).

Horowitz, J. and K. McConnell (2002), "A review of WTA/WTP studies", *Journal of Environmental Economics and Management*, Vol. 44, pp. 426-447.

Hubbel, B.J. (2002), Implementing QALYs in the Analysis of Air Pollution Regulations, US Environmental Protection Agency, Washington, DC, www.epa.gov/ttnecas1/workingpapers/ereqaly.pdf (last accessed on 3 May 2016).

Huijnen, V. et al. (2010), "The global chemistry transport model TM5: description and evaluation of the tropospheric chemistry version 3.0", *Geoscientific Model Development*, Vol. 3(2), pp. 445-473.

Hunt, A. (2011), "Policy Interventions to Address Health Impacts Associated with Air Pollution, Unsafe Water Supply and Sanitation, and Hazardous Chemicals", *OECD Environment Working Papers*, No. 35, OECD Publishing, Paris, http://dx.doi.org/10.1787/5kg9qx8dsx43-en.

IEA (International Energy Agency) (2013), *World Energy Outlook 2013*, IEA, Paris, http://dx.doi.org/10.1787/weo-2013-en.

IEA (2011), *World Energy Outlook 2011*, OECD Publishing, Paris, http://dx.doi.org/10.1787/weo-2011-en.

ILO (International Labour Organisation) (2011), Estimates and Projections of the Economically Active Population: 1990-2020, 6th ed. ILO.

IMF (International Monetary Fund) (2014), World Economic Outlook Database. International Monetary Fund, www.imf.org/external/pubs/ft/weo/2014/01/weodata/index.aspx (last accessed on 3 May 2016).

Izquerdo, J.L. (2003), "The burden of COPD in Spain: results from confronting COPD survey", *Respiratory Medicine*, Mar 97, Suppl C:S61-9.

Leitao, J., R. van Dingenen, S. Rao (2015), LIMITS project Report on spatial emissions downscaling and concentrations for health impacts assessment, www.feem-project.net/limits/docs/limits_d4-2_iiasa.pdf (last accessed on 3 May 2016).

Lim, S.S. et al. (2012), "A comparative risk assessment of burden of disease and injury attributable to 67 risk factors and risk factor clusters in 21 regions, 1990-2010: A systematic analysis for the Global Burden of Disease Study 2010", *The Lancet*, Vol. 380, pp. 2224-60.

Lindhjem, H. and S. Navrud (2008), "How Reliable are Meta-Analyses for International Benefit Transfer?", *Ecological Economics*, 66(2-3); 425-435.

Jones, J. et al. (2003), "The DSSAT cropping system model", *European Journal of Agronomy*, Vol. 18, pp. 235-265.

Kriegler, E. et al. (2014), "What does the 2°C target imply for a global climate agreement in 2020? The LIMITS study on Durban Platform scenarios", *Climate Change Economics*, Vol. 04(4).

Krol, M. et al. (2005), "The two-way nested global chemistry-transport zoom model TM5: algorithm and application", *Atmospheric Chemistry and Physics*, Vol. 4, pp. 3975-4018.

Mayeres, I. and D. van Regemorter (2008), "Modelling the health related benefits of environmental policies and their feedback effects: a CGE analysis for the EU countries with GEM-E3", *The Energy Journal*, Vol. 29(1), pp. 135-150.

Mills, G. et al. (2007), "A synthesis of AOT40-based response functions and critical levels of ozone for agricultural and horticultural crops", *Atmospheric Environment*, Vol. 41, pp. 2630-2643.

OECD (2015), *The Economic Consequences of Climate Change*, OECD Publishing, Paris, http://dx.doi.org/10.1787/9789264235410-en.

OECD (2014a), *The Cost of Air Pollution: Health Impacts of Road Transport*, OECD Publishing, Paris, http://dx.doi.org/10.1787/9789264210448-en.

OECD (2014b), *Towards Green Growth in Southeast Asia*, OECD Green Growth Studies, OECD Publishing, Paris, http://dx.doi.org/10.1787/9789264224100-en.

OECD (2014c), *OECD Economic Outlook*, OECD Publishing, Paris, http://dx.doi.org/10.1787/eco_outlook-v2014-2-en.

OECD (2012), *Mortality Risk Valuation in Environment, Health and Transport Policies*, OECD Publishing, Paris, http://dx.doi.org/10.1787/9789264130807-en.

OECD (2011), "Labour Force Statistics : Population projections", OECD Employment and Labour Market Statistics (database), http://dx.doi.org/10.1787/data-00538-en (last accessed on 3 May 2016).

OECD (2010), *Valuation of Environment-Related Health Risks for Children*, OECD Publishing, Paris, http://dx.doi.org/10.1787/9789264038042-en.

OECD (2006), *Cost-Benefit Analysis and the Environment: Recent Developments*, OECD Publishing, Paris, http://dx.doi.org/10.1787/9789264010055-en.

Piperno, D. et al. (2003), "The burden of COPD in France: results from the confronting COPD survey", *Respiratory Medicine*, Mar 97, Suppl. C:S33-42.

Rao, S. et al. (2016). A multi-model analysis of the co-benefits of climate change mitigation for global air quality, submitted to Environmental Research Letters.

Rao, S. et al. (2012), "Environmental Modeling and Methods for Estimation of the Global Health Impacts of Air Pollution", *Environmental Modeling and Assessment*, Vol. 17, pp. 613-622.

Ready, R. et al. (2004). "Benefit transfer in Europe: How reliable are transfers between countries?", *Environmental and Resource Economics*, 29, pp. 67-82.

Riahi, K., A. Gruebler, and N. Nakicenovic (2007), "Scenarios of long-term socio-economic and environmental development under climate stabilization", *Technological Forecasting and Social Change,* Vol. 74(7), pp. 887-935.

Rosegrant, M.W. et al. (2012), "International Model for Policy Analysis of Agricultural Commodities and Trade (IMPACT) Model Description", International Food Policy Research Institute, www.ifpri.org/sites/default/files/publications/impactwater2012.pdf (last accessed on 3 May 2016).

Stieb, D.M., et al. (2002), "Economic evaluation of the benefits of reducing acute cardiorespiratory morbidity associated with air pollution", *Environmental Health*, Vol. 1(7).

Sue Wing, I. and K. Fisher-Vanden. (2013), "Confronting the Challenge of Integrated Assessment of Climate Adaptation: A Conceptual Framework", *Climatic Change*, Vol. 117(3), pp. 497-514.

Sue Wing, I. and E. Lanzi (2014), "Integrated Assessment of Climate Change Impacts: Conceptual Frameworks, Modelling Approaches and Research Needs", *OECD Environment Working Papers*, No. 66, OECD Publishing, Paris, http://dx.doi.org/10.1787/5jz2qcjsrvzx-en.

UN (United Nations) (2015), World Population Prospects: The 2015 Revision, Key Findings and Advance Tables. Working Paper No. ESA/P/WP.241.

UNEP (United Nations Environment Programme) (2011), Near-term Climate Protection and Clean Air Benefits: Actions for Controlling Short-Lived Climate Forcers, United Nations Environment Programme (UNEP), Nairobi, Kenya.

Van Dingenen, R. et al. (2009), "The global impact of ozone on agricultural crop yields under current and future air quality legislation", *Atmospheric Environment*, Vol. 43(3), pp. 604-618.

Vrontisi, Z. et al. (2016), "Economic impacts of EU clean air policies assessed in a CGE framework", *Environmental Science and Policy*, Vol. 55, pp. 54-64.

Wagner, F. and Amann, M. (2009), GAINS (Greenhouse gas – air pollution interactions and synergies): analysis of the proposals for GHG reductions in 2020 made by UNFCCC Annex 1 countries by mid-August 2009: IIASA, Laxenburg, Austria.

Wagner, F. et al. (2007), The GAINS Optimization Module as of 1 February 2007, Interim Report IR-07-004, IIASA, Laxenburg, Austria.

World Bank (2015), *World Development Indicators*, Washington, DC., http://data.worldbank.org/indicator (last accessed on 3 May 2016).

World Bank (2014), Doing Business 2015: Going Beyond Efficiency, The World Bank, Washington, DC. http://dx.doi.org/10.1596/978-1-4648-0351-2.

WHO (World Health Organisation) (2014), Burden of disease from Household Air Pollution for 2012. The World Health Organization, Geneva 2014.

WHO (2013), Health risks of air pollution in Europe – HRAPIE project. Recommendations for concentration-response functions for cost-benefit analysis of particulate matter, ozone and nitrogen dioxide, World Health Organization, Regional Office for Europe, Bonn, Germany. www.euro.who.int/en/health-topics/environment-and-health/air-quality/publications/2013/health-risks-of-air-pollution-in-europe-hrapie-project-recommendations-for-concentrationresponse-functions-for-costbenefit-analysis-of-particulate-matter,-ozone-and-nitrogen-dioxide (last accessed on 3 May 2016).

Chapter 3

Projections of economic growth and impacts of outdoor air pollution

This chapter outlines the main socioeconomic trends that are projected to emerge in absence of environmental policies other than those that are already in place. It presents the projections of the air pollutant emissions as linked to the economic projections of the ENV-Linkages model. The chapter also presents results on the concentrations of key pollutants that are the drivers of impacts on health and crop yields. Finally, it presents results on the biophysical impacts related to premature deaths, increasing cases of illnesses, and changes in crop yields.

3.1. Trends in economic activity and growth

The projections of future economic activity are based on a modelling framework, the OECD's ENV-Linkages model (Chateau et al., 2014), which provides trends in sectoral and regional economic activity. These projections of GDP and other economic indicators are driven by a multitude of factors, including assumptions on so-called megatrends, such as developments in demography and technology. These megatrends are country-specific. For example, the age structure in the People's Republic of China (henceforth "China") and in India are different: aging will become a major force in China in the coming decades, while India has a much younger population. Similarly, while the average annual growth rate of technological progress is currently highest in the emerging economies, such as China, India and Indonesia, growth rates in these countries are projected to decline, while they are projected to increase in many developing countries.

The regional projections of GDP indicate that global economic activity will continue growing in the coming decades. While long-run global economic growth rates are gradually declining, Figure 3.1 shows that GDP levels in the projection without economic feedbacks from air pollution are still projected to increase significantly over time. The largest growth is projected to be outside the OECD region, especially in Asia and Africa, where a huge economic growth potential exists. The share of the OECD in the world economy is projected to decline from 64% in 2010 to 38% in 2060. These projections are fully aligned with the OECD Economic Outlook (OECD, 2014) and include the main effects of the financial crisis as they emerged until 2013. They are also consistent with the central scenario of the OECD@100 report on long-term scenarios (Braconier et al., 2014).

Figure 3.2 shows how the sectoral structure evolves in the regional economies. The shares of the various sectors in OECD economies tend to be relatively stable, with the services sectors accounting for more than half of GDP (i.e. value added). However, there are undoubtedly many fundamental changes at the sub-sectoral level that are not reflected here.

Figure 3.1. **Trend in real GDP, no-feedback projection**

Billions of USD, 2010 PPP exchange rates

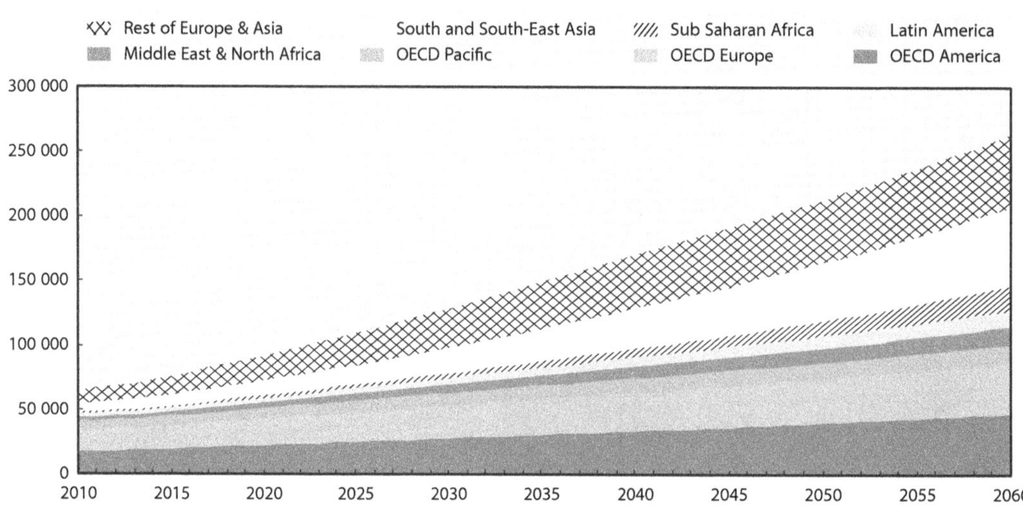

StatLink http://dx.doi.org/10.1787/888933357300

Source: OECD (2014) for OECD countries and ENV-Linkages model for non-OECD economies.

Figure 3.2. **Sectoral composition of GDP by region, no-feedback projection**
Percentage of GDP, 2010, 2035 and 2060

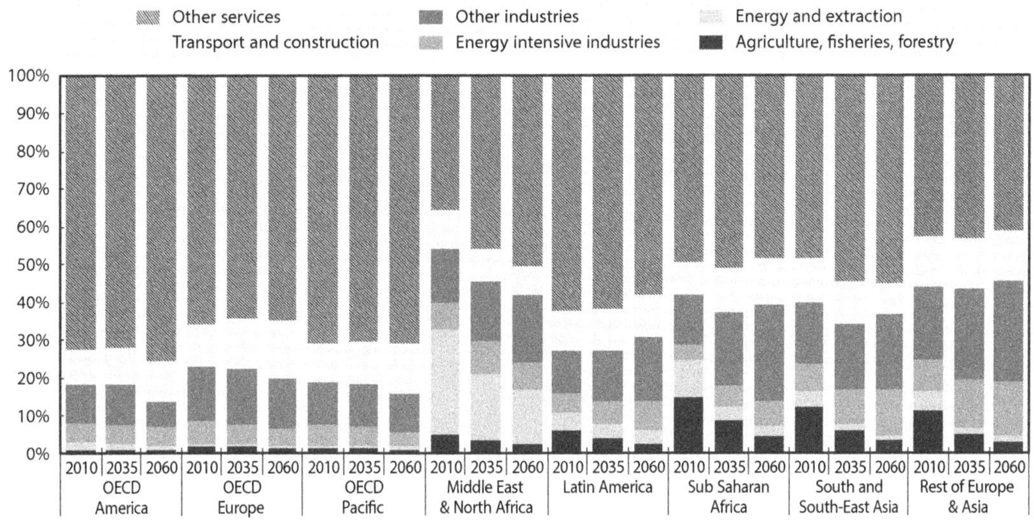

StatLink http://dx.doi.org/10.1787/888933357312

Source: ENV-Linkages model.

The major oil exporters in the Middle East and Northern Africa are projected to gradually diversify their economies and rely less on energy resources. In developing countries the decline of the importance of agriculture is projected to continue strongly. Given the high growth rates in many of these economies, this does not mean an absolute decline of agricultural production, but rather an industrialisation process, and, in many cases, a strong increase in services. Energy and extraction increases especially in the South and South-East Asia and Rest of Europe and Asia regions, reflecting a higher reliance on fossil fuels and a strong increase in electricity use. This has significant consequences for emissions of air pollutants.

3.2. Projections of air pollutant emissions

For most air pollutants, emissions are projected to increase in the coming decades, as illustrated in Figure 3.3. Rising emissions reflect the underlying baseline assumptions on economic growth, as presented in Section 3.1. With increasing GDP and energy demand, especially in some fast growing economies such as India and China, emissions of air pollutants rise at global level.

Emissions of nitrogen oxides (NO_x) and ammonia (NH_3) are projected to have a particularly strong increase, with NO_x emissions almost doubling by 2060. These large changes are due to the projected increase in the demand for agricultural products and energy (incl. transport and power generation) and a rather limited control of NO_x emissions from power plants and industrial boilers in the developing world. Emissions of all other pollutants also increase with the exception of organic carbon (OC). The slight emission decrease for OC corresponds to lower emissions from energy demand from households, which reflects technology improvements in energy efficiency, the use of cleaner fuels, and the switch from biomass in open fire to cleaner energy sources including LPG, ethanol, or enhanced cooking stoves. Interestingly, emissions of sulphur dioxide (SO_2) are projected to initially decrease but increase again after 2030. The initial decline is due to current policies that require flue gas desulphurisation even in several developing countries (primarily in the power sector), but is later offset by the continuing increase in energy demand, which eventually leads to higher emissions.

Figure 3.3. **Emission projections over time**

Index with respect to 2010

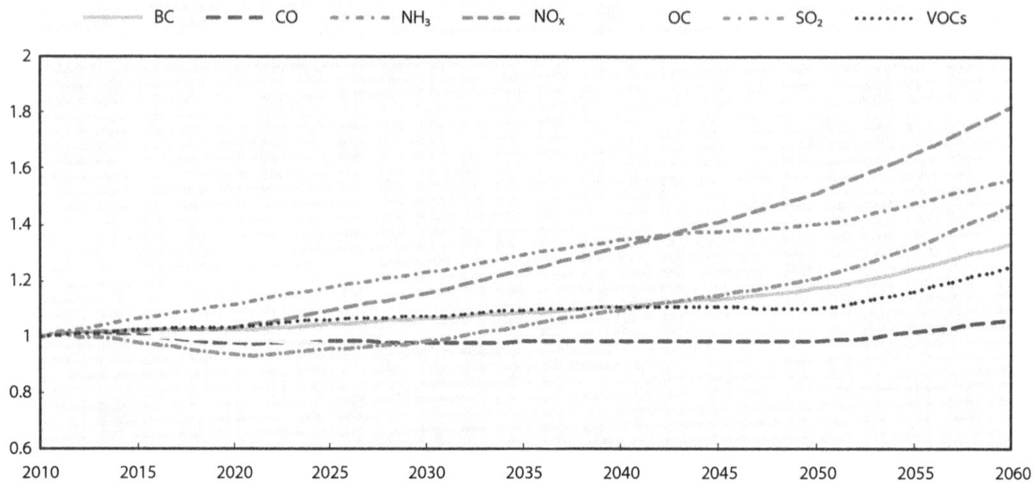

StatLink http://dx.doi.org/10.1787/888933357324

Source: ENV-Linkages model, based on projections of emission factors from the GAINS model.

Each gas has a unique profile of different emission sources, as illustrated in Figure 3.4. The emission sources considered are grouped into energy demand from industrial sectors, from residential and commercial services and from transport, energy supply, land use, and emissions from waste, wastewater treatment, and solvents.

With the exception of emissions of NH_3, which are mostly caused by livestock production and land use with associated application of manures and mineral fertilisers, the energy sector is the main source of air pollutant emissions (IEA, 2016). More specifically, the main emission sources are linked to combustion processes and energy use.

Figure 3.4. **Sectoral shares of emissions**

Percentage of total emissions

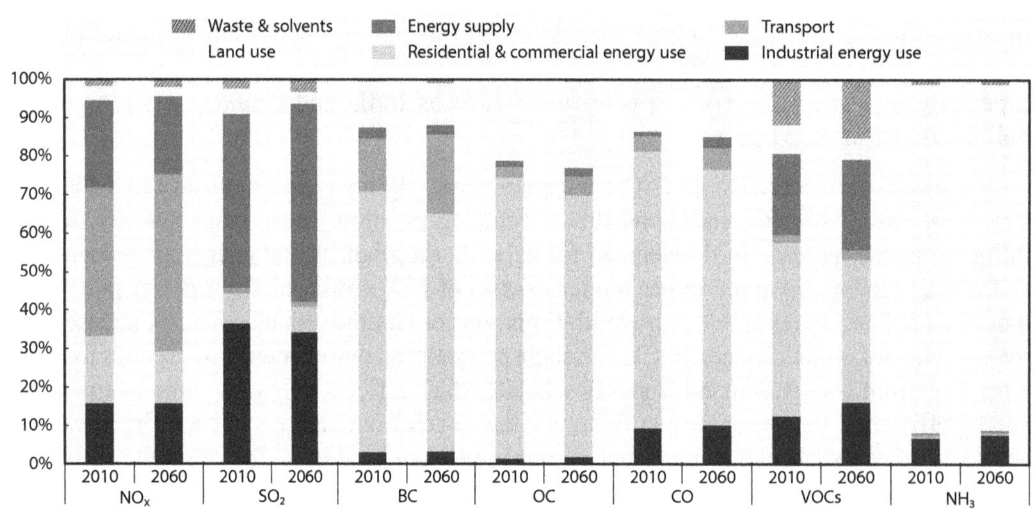

StatLink http://dx.doi.org/10.1787/888933357334

Source: ENV-Linkages model, based on projections of emission factors from the GAINS model.

Figure 3.5. **Emissions by region and by pollutant**

Megatonnes

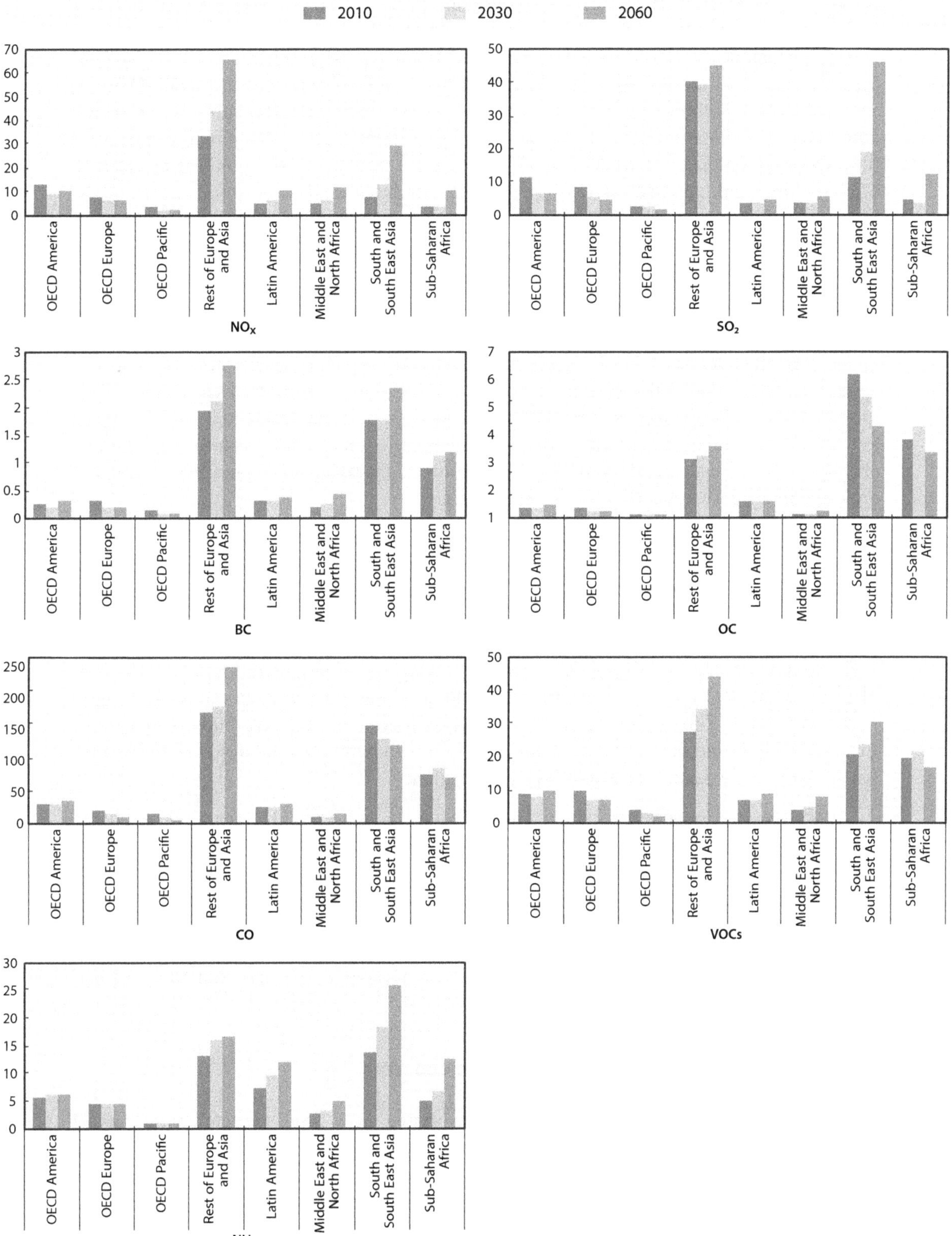

StatLink http://dx.doi.org/10.1787/888933357344

Source: ENV-Linkages model, based on projections of emission factors from the GAINS model.

Most emissions of NO_x and SO_2 come from combustion processes in respectively transport and energy supply (power generation). However, the majority of emissions of NO_x originate from transport and industrial sources, whereas emissions of SO_2 are almost completely from industrial sources including power generation. In the United States, for instance, in 2010 a large source of NO_x emissions (around 33%) was road (and rail) transport and around 70% of SO_2 emissions were from coal power plants. Primary sources of black carbon (BC) and OC emissions are caused by the transport sector, which dominates emissions in OECD countries, and by residential and commercial use of solid fuels (cooking and heating in developing countries). Locally, informal industries, such as brick making, and land use, such as open burning of biomass, can be significant sources. Emissions of carbon monoxide (CO) and volatile organic componds (VOCs) stem from all sources, with only a small contribution from energy supply. The largest share of VOCs emissions is from waste and solvents.

While the sources of emissions are largely unchanged over time, they do not all grow at the same rate over time. The contribution of emissions from industrial sources is projected to increase for all pollutants. The contribution of emissions from land use is by contrast projected to decrease.[1] The contribution of emissions from residential and commercial services for CO is projected to remain relatively stable. Emission reductions from the residential sectors, thanks to technological improvements, are offset by higher emissions from transport and industrial energy demand. Finally, emissions from other sources, including waste and solvent use, are projected to increase, especially for OC and VOCs.

There are large differences among countries and regions in emissions of the different pollutants, as illustrated in Figure 3.5. NO_x emissions are particularly high in the Rest of Europe and Asia region (which includes China) but also high in South and South East Asia. Emissions of SO_2 are also high in the Rest of Europe and Asia region for the reference year 2010. However, by 2060 emissions in the South and South East Asia region become equally high. This is mostly due to emissions rapidly rising in India and Indonesia. Emissions of BC and OC, CO and VOCs are highest in Rest of Europe and Asia, South and South East Asia and in Sub-Saharan Africa. Finally, emissions of NH_3 are highest in Rest of Europe and Asia and in South and South East Asia, although they are projected to increase particularly in the South and South East Asia region.

Emissions are generally projected to increase in non-OECD economies, with the highest increases taking place in the South and South East Asia region. The exception to this is emissions of OC and CO that decline in South and South East Asia and Sub-Saharan Africa. This is mostly thanks to improvement in the residential sectors, i.e. access to cleaner energy for households, linked to general megatrends, including urbanisation and electrification. Emissions from OECD countries tend to be stable or to slightly decline, although the projections show a small increase in emissions of all gases but NO_x and SO_2 in the OECD America region.

3.3. Projections of particulate matter and ozone concentrations

With emissions of air pollutants generally rising over time, the concentrations of $PM_{2.5}$ and ozone are also projected to increase in most regions, although, as discussed in Chapter 2, climatic conditions and several other factors influence concentrations. The maps in Figure 3.6 illustrate the annual average of anthropogenic $PM_{2.5}$ concentrations in the reference year (2010) as well as in the projected years 2030 and 2060 (maps for overall emissions, including the natural components of dust and sea salt, are presented in the right panels). In the reference year, $PM_{2.5}$ concentrations are highest in South and East Asia, and particularly in China and India. They are also high in some areas of North America, Europe and Africa. According to the projections, the average concentrations will increase significantly in South and East Asia, as well in some areas of Africa. Concentrations are projected to slightly decrease in North America and Europe.

The projections of average $PM_{2.5}$ concentrations show that several areas are already well above the reference levels recommended by the WHO air quality guidelines (WHO, 2006) (see Box 3.1). The WHO guidelines recommend annual mean concentrations below 10 μg/m³ but also specify interim targets that reflect achievable levels with abatement measures. The highest interim target is set at 35 μg/m³ and is estimated to correspond to 15% higher long-term mortality risk relative to the recommended target.

The WHO guideline level of 10 μg/m³ should not be interpreted as a cut-off point below which there are no health impacts. The epidemiological literature is not yet grounded on a reached consensus on what happens at low concentration levels. Recent literature, e.g. Shi et al. (2016), suggests that there is no lower cut-off level regarding impacts, and even concentration levels below 10 μg/m³ may lead to health impacts. The calculations of the health impacts in this report do not include a cut-off point and allow for impacts also at lower concentration levels. Nevertheless, the WHO guidelines provide an insightful reference point to shed light on the severity of the outdoor air pollution problem.

Box 3.1. **WHO global air quality guidelines**

The WHO global air quality guidelines (WHO, 2006), specify target levels for concentrations of particulate matter (both PM_{10} and $PM_{2.5}$) and ozone concerning the health impacts associated with each of the targets. The guidelines are useful in identifying levels of concentrations that do not lead to low effects on human health. Table 3.1 summarises the characteristics of the guidelines and targets for $PM_{2.5}$ and ozone.

Table 3.1. **Targets specified in the WHO air quality guidelines**

Targets		Basis for selected level
$PM_{2.5}$ (Annual mean $PM_{2.5}$, μg/m³)		
Interim target 1	35	About a 15% higher long-term mortality risk relative to guideline level.
Interim target 2	25	In addition to other health benefits, these levels lower the risk of premature mortality by 2-11% relative to the Interim target 1 level.
Interim target 3	15	In addition to other health benefits, these levels reduce the mortality risk by 2-11% relative to the interim target 2.
Air quality guideline	10	These are the lowest levels at which mortality due to long-term exposure to $PM_{2.5}$ has been shown to increase with more than 95% confidence.
Ozone (Daily maximum 8-hour mean, μg/m³)		
High levels	240	Significant health effects; substantial proportion of vulnerable populations affected.
Interim target 1	160	Important health effects; does not provide adequate protection of public health.
Air quality guideline	100	Provides adequate protection of public health, though some health effects may occur below this level.

Source: WHO (2006).

Figure 3.6. **Particulate matter concentrations**

Annual average total $PM_{2.5}$; anthropogenic on left panels and total on right panels, µg/m³

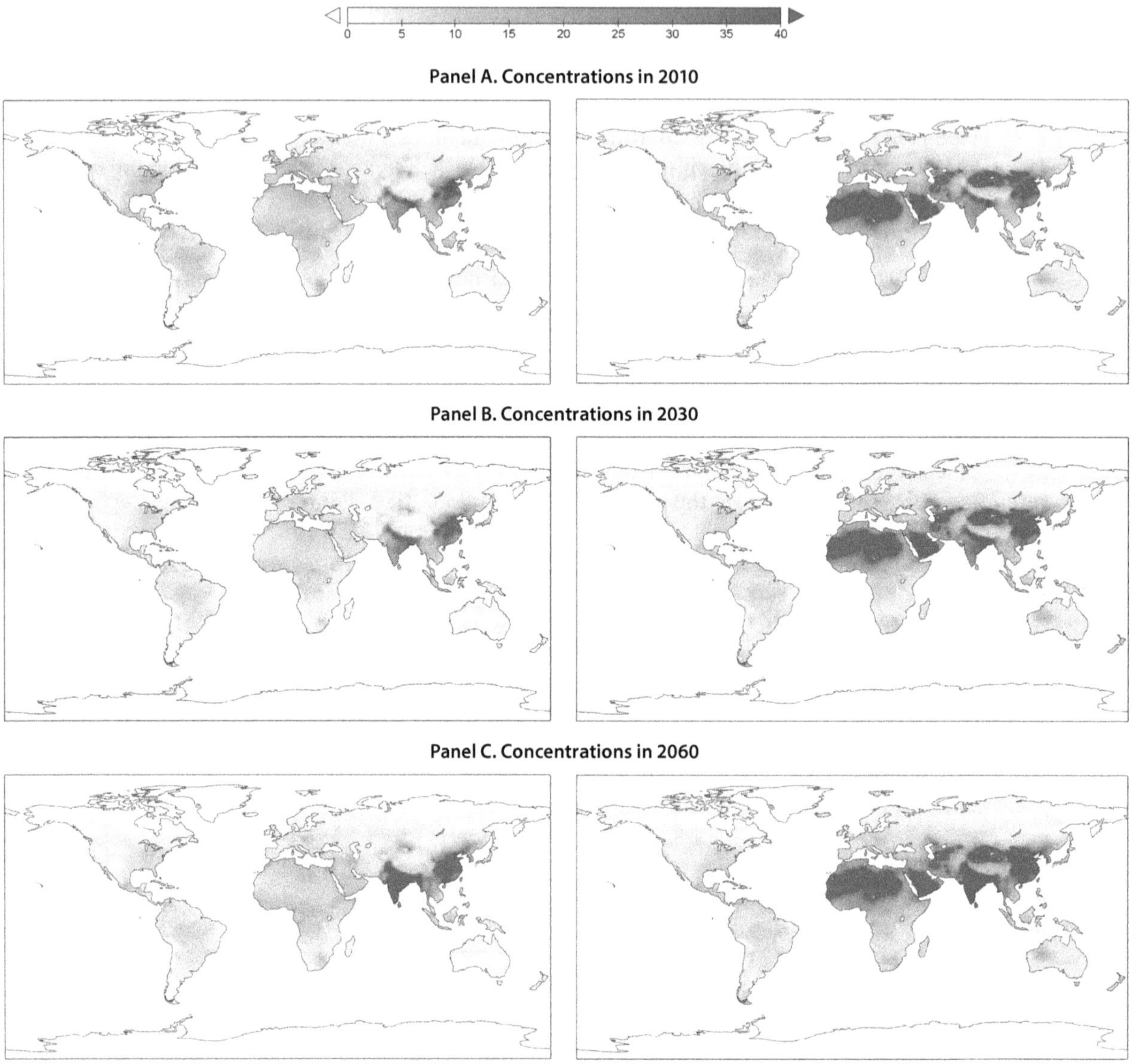

Note: The maps are based on concentrations specified at a 1°×1° resolution.

Source: TM5-FASST model, based on projections of emissions from the ENV-Linkages model.

As illustrated in Figure 3.6, several world regions, and especially China and India, were already above the highest interim target in 2010 and are projected to reach even higher levels by 2060. While the maps in Figure 3.6 show lighter colours for OECD regions, these levels are above the recommended WHO guidelines in most areas, implying that there are still strong impacts on human health and the environment.

Less than 4 people out of 10 around the globe live in areas that respect the levels of $PM_{2.5}$ concentrations recommended by the WHO Air quality Guidelines (10 µg/m³). Even below this threshold, there may still be impacts on human health. The population exposure also changes over time. The percentage of population exposed to annual mean concentrations of $PM_{2.5}$ higher than 35 µg/m³ is estimated to be 15% in 2010, while it is projected to increase to 30% by 2060. The increase is even higher in China and India where emissions and concentrations increase the most. In China the percentage of population exposed to annual mean concentrations of $PM_{2.5}$ higher than 35 µg/m³ is projected to

Figure 3.7. **Ozone concentrations**

Maximal 6-month mean of daily maximal hourly ozone, M6M, in ppb

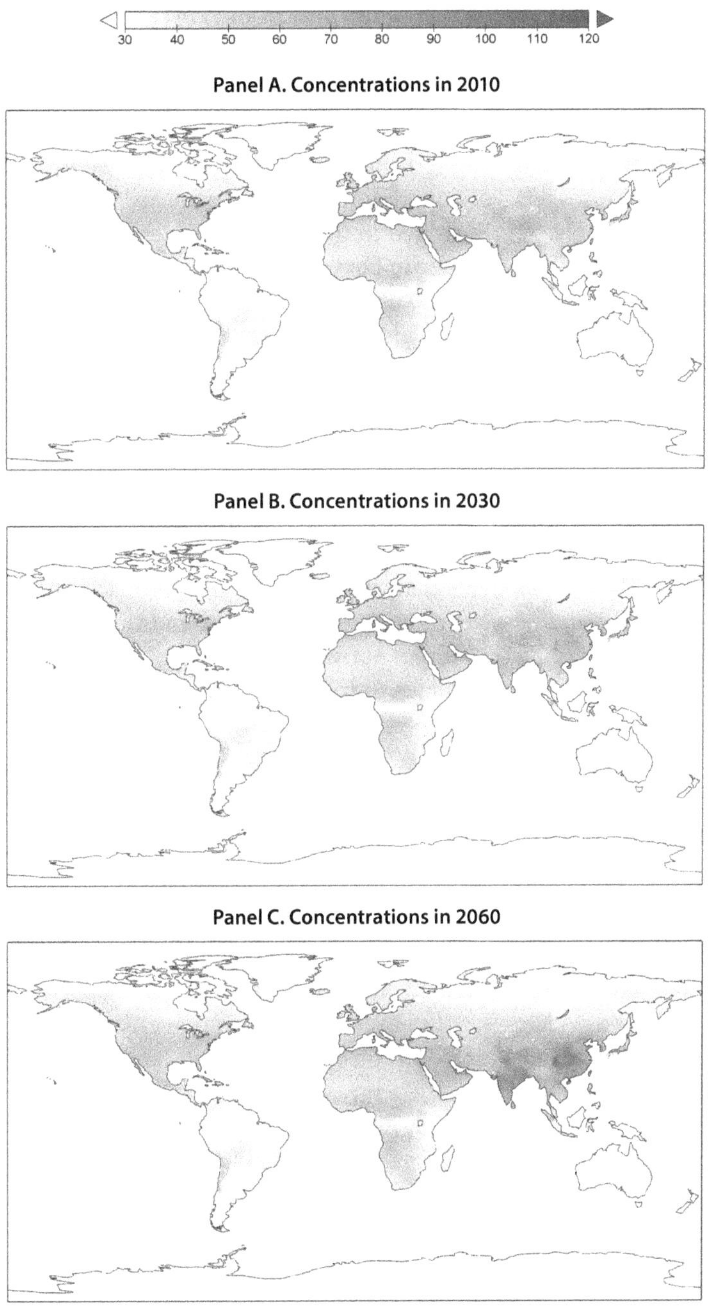

Note: The maps are based on concentrations specified at a 1°×1° resolution.

Source: TM5-FASST model, based on projections of emissions from the ENV-Linkages model.

increase from 40% in 2010 to 65% in 2060 while in India from 15% in 2010 to 60% in 2060. While in China the percentage is already currently very high, the worsening of the population exposure is particularly strong in India.

Average concentrations of ground level ozone are presented in Figure 3.7. They are particularly high in parts of Asia (not least Korea), the Middle East and the Mediterranean, but they also exceed air quality guidelines in many other OECD and non-OECD regions. These areas are most affected not only in the reference year but also in the projections at both 2030 and 2060. While there are hardly any changes by 2030, there are some more significant changes by 2060. According to the projections, by 2060, some areas, including parts of the Middle East and Asia (including China and India) could reach very high levels of concentrations (above 120 ppb maximal 6-month mean of daily maximal hourly ozone).

For ozone concentrations, the WHO guidelines recommend levels below 100 ($\mu g/m^3$) daily maximum 8-hour mean, with the highest interim target set at 240 ($\mu g/m^3$) daily maximum 8-hour mean. These are respectively approximately equivalent to 50 and 120 ppb 6-month mean daily maximum levels. As illustrated in the maps, considering average concentrations there are no areas above the highest interim target in 2010. However, by 2060, the projections show that several areas will reach levels above the interim target, especially in China and India. Such high concentrations will lead to significant health effects and environmental impacts, including reductions in crop yields that will affect agricultural output.

For the scope of this report, concentrations of pollutants are only an intermediate step for the calculations of the economic consequences of air pollution. Nevertheless, the average numbers presented are themselves an indicator of the severity of the air pollution problem. High average numbers mean that in many areas – and especially in large cities – air pollution levels are permanently above recommended levels and that there are several days per year where they reach levels that are extremely dangerous for human health. This has already happened in the past years in several cities around the globe, affecting health but also leading to restrictions on human activities. This type of situation is projected to increase in the absence of further policies to reduce air pollutant emissions.

3.4. Projections of the impacts of outdoor air pollution on health and agriculture

3.4.1. Premature deaths

The number of premature deaths due to outdoor air pollution have already been estimated to be high in recent years (see e.g. Lim et al., 2012 and Forouzanfar, 2015), with elderly people and children being most affected (WHO, 2014). The fundamental issue in estimating the number of premature deaths due to air pollution is the shape of the concentration-response function over a wide range of observed concentrations. For the base year 2010, the calculations of premature deaths are based on the Global Burden of Disease work reported by Forouzanfar et al. (2015) for $PM_{2.5}$ and Lim et al. (2012) for ozone. For future projections, the concentration-response function for $PM_{2.5}$ in particular becomes more uncertain as the population-weighted concentrations of $PM_{2.5}$ become much higher in some countries. To reflect this uncertainty two different functions are used for $PM_{2.5}$: *(i)* a linear function showing a simple linear relationship between concentrations and the number of premature deaths adjusted for changes in mortality rates, and *(ii)* a non-linear function, which considers that the incremental number of deaths decreases as concentrations become higher. Annex C outlines in more detail the two different formulations of the concentration-response function.

According to the calculations, premature deaths caused by outdoor air pollution in the reference year 2010 amounted to almost 3 million people globally (in line with the results of Forouzanfar et al., 2015). Premature deaths from outdoor air pollution are projected to reach a global total of 6 to 9 million people in 2060 (considering a non-linear and a linear concentration-response function respectively). This large increase is not only due to higher concentrations of $PM_{2.5}$ and O_3, but also to an increasing and aging population and to urbanisation (which also leads to higher exposure).

High concentrations of $PM_{2.5}$ account for most of the premature deaths. In 2010, PM is linked to around 95% of premature deaths from air pollution at the global level. The contribution of PM to mortality varies across regions. This fraction is lowest in India (89%) and highest in regions such as Canada where PM is responsible for almost all premature

Table 3.2. **Premature deaths from exposure to particulate matter and ozone**

Number of premature deaths caused by outdoor air pollution, thousands of people

		2010	2030		2060	
			Non-linear	Linear	Non-linear	Linear
OECD America	Canada	8	10	10	13	14
	Chile	3	4	4	7	6
	Mexico	14	21	21	42	42
	USA	93	92	99	122	128
OECD Europe	EU large 4	111	97	98	89	95
	Other OECD EU	90	87	84	99	97
	Other OECD	28	37	35	65	64
OECD Pacific	Aus. & New Z.	2	2	3	3	4
	Japan	60	78	76	77	80
	Korea	17	31	30	52	54
Rest of Europe & Asia	China	905	1 374	1 492	2 065	2 711
	Non-OECD EU	33	26	25	23	22
	Russia	119	106	107	93	93
	Caspian region	44	69	69	111	116
	Other Europe	74	57	56	49	49
Latin America	Brazil	36	48	48	73	73
	Other Lat. Am.	38	52	53	87	87
Middle East & North Africa	Middle East	52	85	95	191	229
	North Africa	52	65	62	107	112
South and South-East Asia	ASEAN 9	102	152	155	286	343
	Indonesia	57	80	81	113	116
	India	613	788	926	1 553	3 351
	Other Asia	202	253	253	509	811
Sub-Saharan Africa	South Africa	12	8	9	11	11
	Other Africa	167	178	180	323	334
OECD		*428*	*459*	*460*	*569*	*584*
Non-OECD		*2 505*	*3 339*	*3 610*	*5 593*	*8 459*
World		*2 933*	*3 799*	*4 070*	*6 162*	*9 043*

StatLink http://dx.doi.org/10.1787/888933357514

Note: Due to the curvature of the functions and rounding, the effects of the non-linear projection can in some cases be reported to be slightly higher than the linear projection; this only affects the results for low and modest concentration levels.

deaths linked to outdoor air pollution. Whilst PM accounts for the highest share of deaths, mortality due to ozone is projected to increase over time as ozone concentrations become higher and more dangerous for human health. By 2060, premature deaths due to ozone are projected to increase to 7-10% of the total. In India, they could account for up to 20%.

The number of premature deaths is unequally distributed across the world. As illustrated in Table 3.2, the highest number of deaths takes place in non-OECD economies and particularly in China and India. These regions also experience the highest increase in the number of premature deaths to 2060. China's premature deaths account for 31% of the global total in 2010 and for 30-34% in 2060. While China's share of premature deaths is rather stable over time, premature deaths in India increase substantially over time and increase from 21% of the global total in 2010 to 27-35% in 2060. A smaller increase is projected in OECD countries, with the number of premature deaths increasing from around 430 thousand people in 2010 to around 570-580 thousand in 2060. The share of premature deaths caused by outdoor air pollution in OECD countries decreases over time (from 15% of the global total in 2010 to 6-9% in 2060). In particular the share of premature deaths of the United States decreases from 3% of the global total in 2010 to 1-2% in 2060, and from 8% in 2010 to 2-3% for the EU.

The range of projected results in 2060 is larger in some regions than in others. For regions where the increase in concentrations is limited, there is hardly any difference between the results obtained with the two alterative functions. For regions with high increases in concentrations, such as India and China but also South and South East Asia, the range can be quite large. The projected concentrations are larger with the linear function as it considers that premature deaths will continue increasing strongly even with high concentrations of PM.

As already discussed, the increasing number of deaths is partly due to increasing populations, which also lead to a higher number of people being exposed to air pollution. Some of the most affected areas are also highly populated. Nevertheless, even considering the number of premature deaths per million people (Figure 3.8), India and China are projected to have an extremely high number of deaths. Africa, Oceania and Latin America are by contrast the regions with the lowest number of premature deaths per million people.

Figure 3.8. **Premature deaths from exposure to particulate matter and ozone**

Number of deaths caused by outdoor air pollution per year per million people

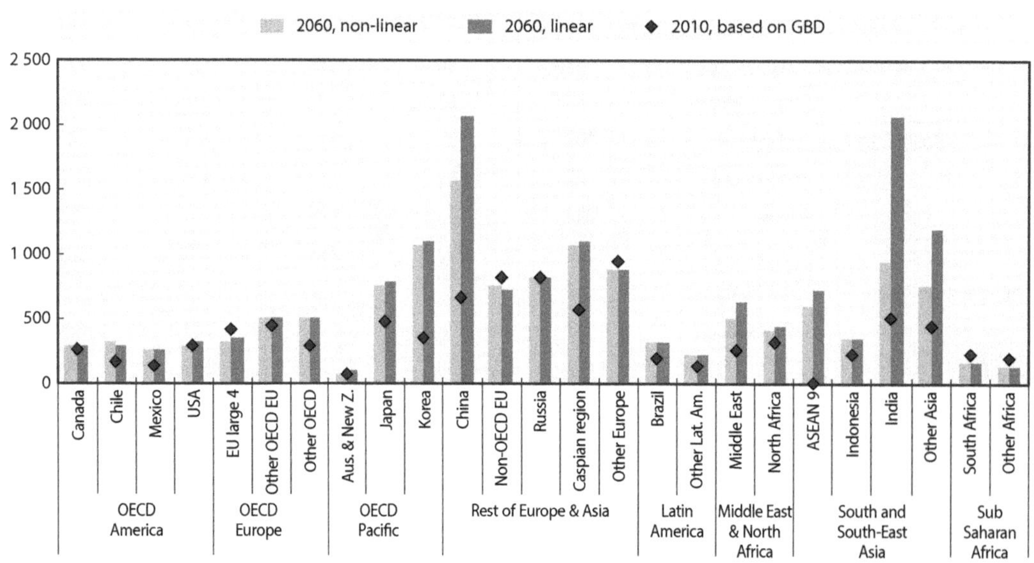

StatLink http://dx.doi.org/10.1787/888933357356

3.4.2. Illness

As previously discussed, increasing concentrations of $PM_{2.5}$ and ozone will also lead to a higher number of cases of illness, which will imply more hospital admissions, health expenditures and sick or restricted activity days, which lead to labour productivity losses.

Table 3.3 presents an overview of the health impacts at the global level. The number of cases of bronchitis is projected to increase substantially going from 12 to 36 million new cases per year for children aged 6 to 12, and from 3.5 to 10 million cases for adults.[2] Children are also affected by asthma, with an increasing number of asthma symptom days for children of age 5 to 19.

These increasing cases of illnesses have been translated into an equivalent number of hospital admissions and then into their corresponding healthcare costs (see Annex C). According to the calculations, hospital admissions are projected to increase from 3.6 in 2010 to 11 million in 2060.

The additional cases of illnesses also lead to an impact on normal work activities. In 2060, lost working time at the global level will be of the order of 3.75 billion days. But there will also be an increasing number of (minor) restricted activity days.

While Table 3.3 presents results at the global level, there are regional differences, which reflect the levels of concentrations of pollutants, the exposure in the different areas and the demographic characteristics of the population. The additional health costs associated with these impacts also vary across the world, reflecting the differences in the capacity and financing of the health systems and the average costs of hospital admissions.

As discussed in Section 2.6, the number of additional cases of illness and of hospital admission is used to calculate overall health expenditures using the established unit values for each endpoint. Lost working days are used to calculate labour productivity changes, as described in Section 2.7. Overall additional health expenditures and changes in labour productivity are then used as inputs in the ENV-Linkages model to calculate the related market costs. Results on market costs are presented in Chapter 4.

The results for additional cases of illness, hospital admissions and (minor) restricted activity days are also used to calculate the welfare costs (e.g. related to pain and suffering) by multiplying results for each endpoint with the appropriate unit value, as explained in Section 2.8. These non-market costs will be presented in Chapter 5.

Table 3.3. **Health impacts at global level**

	2010	2060
Respiratory diseases (million number of cases)		
Bronchitis in children aged 6 to 12	12	36
Chronic bronchitis (adults, cases)	3.5	10
Asthma symptom days (million number of days)		
Asthma symptom days (children aged 5 to 19)	118	360
Healthcare costs (million number of admissions)		
Hospital admissions	3.6	11
Restricted activity days (million number of days)		
Lost working days	1 240	3 750
Restricted activity days	4 930	14 900
Minor restricted activity days (asthma symptom days)	630	2 580

StatLink http://dx.doi.org/10.1787/888933357525

3.4.3. Agricultural yield impacts

High levels of concentration of pollutants, and particularly of ozone, reduce crop yields and thus affect agricultural productivity. Figure 3.9 presents the crop yield changes by region for 2060, expressed as a percentage change from the no-feedback crop yield projections. The graph presents the full range across different crops, and, separately, impacts on rice and wheat.

According to the TM5-FASST calculations, and in line with the larger literature (e.g. Mills et al., 2007; Chuwah et al., 2015), crop yields are negatively affected in all regions, with big differences between regions and crops. In many regions, wheat and oil seeds are more affected than the other crops, with high losses in several OECD countries, including Japan, Korea and the USA for oilseeds, and China and Other Europe for wheat.

In some regions the effects of outdoor air pollution on crop yields are small. For instance, Chile, the Other OECD and Other Africa regions, Australia and New Zealand, and Indonesia have much smaller impacts than the other regions. The impacts on crop yields are included in the ENV-Linkages model for the assessment of market costs. Results are presented in Chapter 5. The regional differences illustrated can lead to changes in competitiveness so that regions that are less affected can even have economic benefits.

Figure 3.9. **Impacts of outdoor air pollution on crop yields**

Percentage change w.r.t. no-feedback projection, 2060

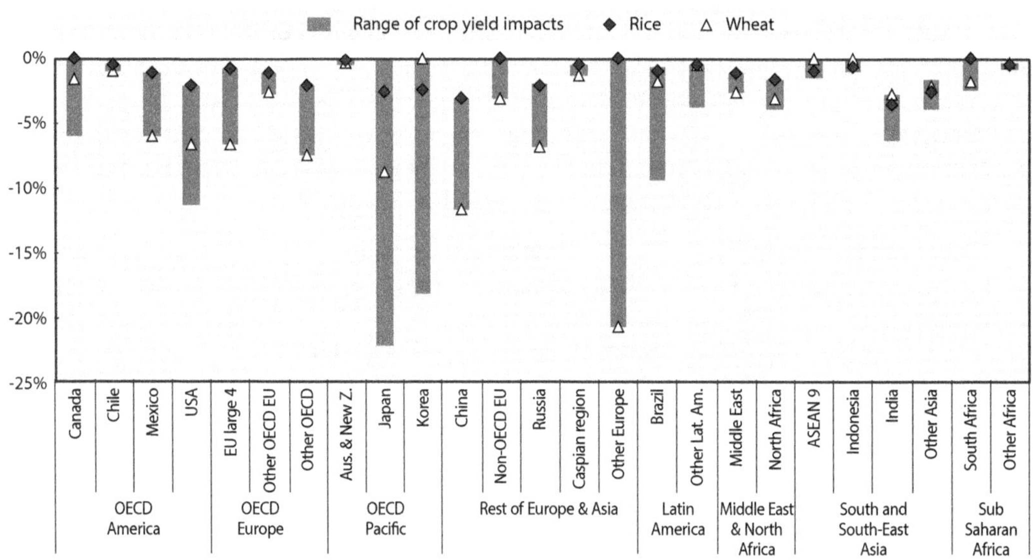

StatLink http://dx.doi.org/10.1787/888933357364

Source: Own calculations, based on the TM5-FASST model and Mills et al. (2007).

Notes

1. This may in part be due to the underestimation of emissions from forest, savannah and agricultural burning.

2. Childhood bronchitis and adult bronchitis persist for different periods. The illness in children typically lasts for only about 2 weeks, whereas in adults, bronchitis may be permanent once initiated.

References

Braconier, H., G. Nicoletti and B. Westmore (2014), "Policy Challenges for the Next 50 Years", *OECD Economic Policy Papers*, No. 9, OECD Publishing, Paris, http://dx.doi.org/10.1787/5jz18gs5fckf-en.

Chateau, J., R. Dellink and E. Lanzi (2014), "An Overview of the OECD ENV-Linkages Model: Version 3", *OECD Environment Working Papers*, No. 65, OECD Publishing, Paris, http://dx.doi.org/10.1787/5jz2qck2b2vd-en.

Chuwah, C. et al. (2015), "Global impacts of surface ozone changes on crop yields and land use", *Atmospheric Environment*, Vol. 106, pp. 11-23.

Forouzanfar, M.H. et al. (2015), "Global, regional, and national comparative risk assessment of 79 behavioural, environmental and occupational, and metabolic risks or clusters of risks in 188 countries, 1990-2013: a systematic analysis for the Global Burden of Disease Study 2013", *The Lancet*, Vol. 386(10010), pp. 2287-2323.

IEA (International Energy Agency) (2016, forthcoming), *World Energy Outlook special report on Energy and Air Pollution*, IEA, Paris.

Lim, S.S. et al. (2012), "A comparative assessment of burden of disease and injury attributable to 67 risk factors and risk clusters in 21 regions, 1990-2010: A systematic analysis for the Global Burden of Disease Study 2010", *The Lancet,* Vol. 380, pp. 2224-60.

Mills, G. et al. (2007), "A synthesis of AOT40-based response functions and critical levels of ozone for agricultural and horticultural crops", *Atmospheric Environment*, Vol. 41, pp. 2630-2643.

OECD (2014), *OECD Economic Outlook*, OECD Publishing, Paris, http://dx.doi.org/10.1787/eco_outlook-v2014-2-en.

Shi, L. et al. (2016), "Low-concentration PM2.5 and mortality: estimating acute and chronic effects in a population-based study", *Environmental Health Perspectives*, Vol. 124(1), pp. 46-52.

WHO (World Health Organisation) (2014), *Burden of disease from Household Air Pollution for 2012*. The World Health Organization, Geneva 2014.

WHO (2006), Air Quality Guidelines: Global Update 2005. www.euro.who.int/__data/assets/pdf_file/0005/78638/E90038.pdf?ua=1 (last accessed on 3 May 2016).

Chapter 4

Consequences of outdoor air pollution for economic growth

This chapter presents the results of the numerical simulations with the ENV-Linkages model on the macroeconomic costs of outdoor air pollution. It first presents the results relative to each impact considered in the report, and then it illustrates results of the impacts as considered together. The focus of this chapter is on market impacts, and macroeconomic costs, but the chapter also investigates regional and sectoral consequences. The results include both direct market impacts, such as those related to changes in crop yields, and indirect impacts, such as those related to the changes in international trade flows due to the regional changes in crop yields.

4.1. Economic consequences of specific market impacts

The market impacts described in Chapter 3 are treated as inputs into the economic modelling framework, ENV-Linkages, to assess how they affect economic activity in the different sectors and regions. Each impact is linked to a specific part of the economic system: lost working days are linked to labour productivity losses, additional health expenditures are linked to increases in demand for healthcare services by both governments and households, and agricultural yield impacts are linked to reduced productivity of agricultural production. The economic consequences are then assessed for the 2015-60 period.

4.1.1. Consequences of the labour productivity impacts

The lost working days related to poor health due to air pollution have a direct effect on labour markets through a reduction of labour productivity, and thus on the contribution of labour to gross domestic product (GDP). Labour supply effects are not included in the central projection, but investigated in an alternative specification in Section 4.3. Panel A of Figure 4.1 presents the change in regional GDP (expressed as deviation from the no-feedback projection) for the year 2060, decomposed into *(i)* the direct effect on labour (the productivity shock), *(ii)* an indirect effect on labour markets (induced effects on wages and the allocation of labour across sectors), *(iii)* an induced effect on capital markets (as capital accumulation adjusts to changes in households' savings) and *(iv)* a change in other components of GDP (including the change in tax revenues and the value added generated by land and natural resources). The direct productivity shock can be labelled as the direct costs of the market impact, whereas the other components together comprise the indirect market costs.

The direct effect of the labour productivity impacts is negative in all regions: air pollution lowers output per worker, and that lowers economic growth. Global GDP in 2016 is 0.1% below the projected level in the case of no feedbacks from outdoor air pollution on the economy. But this productivity shock leads to adjustments throughout the economy (components ii through iv), causing an overall GDP loss in 2060 of 0.4%. For instance, some labour will move from sectors where lower efficiency of labour can be offset by more capital use to sectors where the shock will be managed by employing more people. Demand patterns will also adjust to the changing production costs in the different sectors. But the slowdown in economic activity also induces a negative effect on the total value added generated by labour. As the productivity shock applies to all sectors of the economy, there is little room to accommodate the shock by reallocating labour between sectors. On balance, the indirect effect on labour remuneration as part of GDP is negative in most regions, but smaller than the direct effect (globally less than 0.1%, although the ratio between direct and indirect effect differs by region).[1]

For capital, the effect is negative, and becomes stronger over time. As wage income is reduced owing to the pollution impacts, households respond by reducing their expenditures, including their savings, making less capital available for investment and thus slowing down capital accumulation.[2] Therefore, the negative capital effect is especially large in regions where the income loss from the labour productivity shock is strong, e.g. the People's Republic of China (henceforth "China"), which has a total GDP loss of more than 1%. Interestingly, the capital effect is much smaller in India, even though India, like China, is also projected to be confronted with very large increases in concentrations and significant reductions in labour productivity. The key difference between the two regions is that in the current decade, the marginal propensity to save (and the capital intensity of production) is much smaller in India than in China. Hence, a reduction in income will mostly lead Indian households to reduce consumption, while in China it has a stronger effect on savings. In later decades, when savings rates in India are projected to rise, it will have benefited from the relatively small capital income loss in the first decades.

For the OECD countries, by 2060 the projected GDP losses are substantially smaller than in the big emerging economies. The strongest effects are projected to be in Japan and Korea. Especially Korea is projected to have significant ozone concentrations, almost as high as China and India. For $PM_{2.5}$ (which is dominant in the effect on labour productivity),

Figure 4.1. **Change in GDP from labour productivity impacts, central projection**

Percentage change w.r.t. no-feedback projection

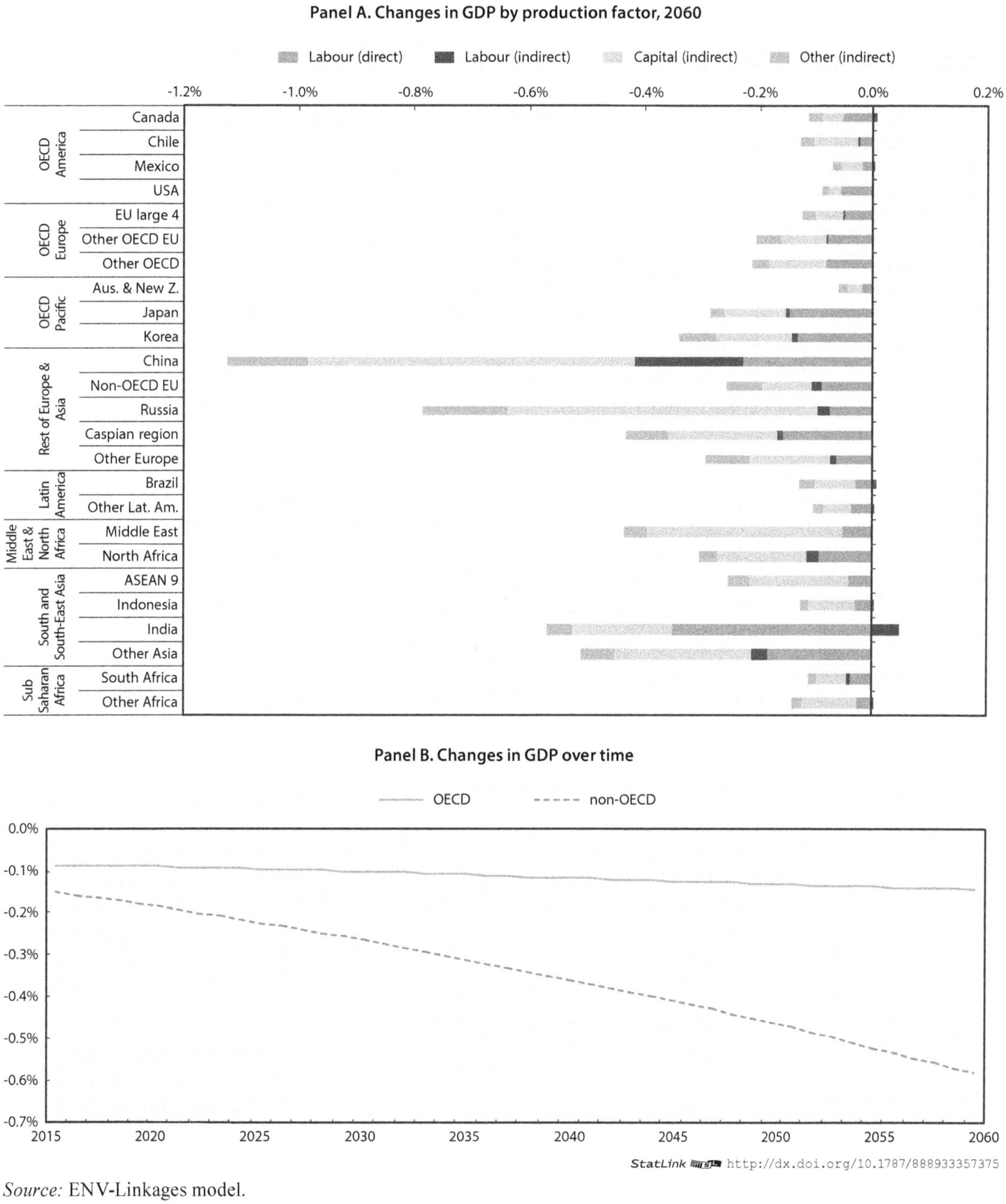

StatLink http://dx.doi.org/10.1787/888933357375

Source: ENV-Linkages model.

the average concentrations in Japan and Korea are, by 2060, significantly lower than in many non-OECD regions, but still higher than in other OECD countries. These higher concentrations translate into higher number of lost working days and thus a stronger labour productivity impact. But as Panel B of Figure 4.1 shows, the differences between the OECD and non-OECD regions become increasingly large in the coming decades. It is primarily the projected economic growth, and associated increases in air pollutant emissions and concentrations, that drive the larger GDP losses in later decades in non-OECD economies.

4.1.2. Consequences of the health expenditure impacts

Health expenditure impacts of air pollution are in a different category from productivity shocks to agriculture or labour. They form a necessary expenditure by governments and households, i.e. an expenditure that is driven by the health impacts of outdoor air pollution rather than by a maximisation of welfare, which leads to reductions of other expenditures, i.e. they affect demand and not productivity.[3] The two possible responses to such a demand shock are to reduce spending on other goods and services (a crowding-out effect), or an increase in total expenditures (an expansion effect). As income is not unlimited, the expansion effect implies that households will limit savings and that governments will need to find a way to finance the expansion, e.g. through increased taxes. In the central projection, both mechanisms are allowed. Households will determine the least-cost mix of reduced consumption of other goods and services and reduced savings; government will not reduce the provision of other public goods and services, but finance the additional health expenditures through an increase in labour taxes (as a proxy for social security payments, following e.g. Vrontisi et al., 2016). Section 4.3 presents an alternative specification where the expansion effect is removed.

As shown in Figure 4.2, the biggest projected changes in health expenditures (and consumption of other non-commercial services, of which health is a part) are in the Rest of Europe and Asia region, which includes China, Russia, the Caspian region and most of Eastern Europe (not least Ukraine).[4] The Caspian and European regions in this group (Non-OECD EU and Other Europe) have a particularly large additional expenditure for pollution-related healthcare relative to other regions. Further, in these regions the share of health expenditures in total expenditures is lower than in OECD regions. Thus, changes related to additional health expenditures are accentuated as – for a given shock – the percentage increase in health spending is higher. Combined, this contributes to *(i)* a significant increase in the non-commercial services sector, and simultaneously *(ii)* a large reduction in consumption of other goods and services (see the corresponding bars in the Figure 4.2).

The effect on GDP follows these sectoral results. On average, the GDP loss in the Rest of Europe and Asia group equals 1.1% by 2060, against 0.4% for the world (not shown). Notably, the effects on the OECD regions is quite small, reflecting that in these economies the additional air-pollution-related health expenditures are on balance a significantly smaller share of total expenditures for both governments and households. In principle, additional expenditures for households have a tendency to lead to larger GDP losses than increased government expenditures, through the induced effect on savings and capital accumulation. Thus, although in many regions the majority of the health expenditures are borne by governments, the majority of the GDP loss can be attributed to additional household expenditures. But as the results for the Rest of Europe and Asia region shows, the minor effects from increased government expenditures only hold at the margin: as soon as the additional expenditures become non-marginal, they will also affect economic growth, not least through the increased tax burden on households that is needed to balance the government budget.[5]

Figure 4.2. **Change in value added and GDP from health expenditure impacts, central projection**

Percentage change w.r.t. no-feedback projection, 2060

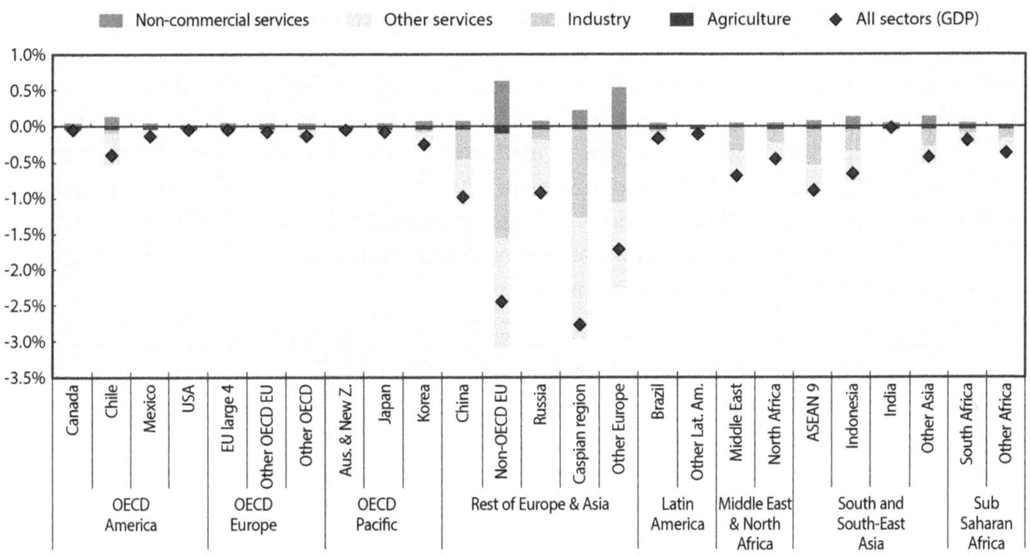

StatLink http://dx.doi.org/10.1787/888933357383

Note: The non-commercial services include health services.

Source: ENV-Linkages model.

4.1.3. Consequences of the agricultural yield impacts

The agricultural yield impacts, as discussed in Section 3.5, lead to a global reduction in the growth of agricultural output over time, i.e. agricultural production declines relative to the no-feedback projection. But given that food is a basic commodity, demand for agricultural products is not very price-elastic and overall crop production does not decline very much (-1.1% in 2060), as shown in Figure 4.3. The lower productivity of agricultural production and the associated increase in unit production costs induce both an intensification and extensification of production: in all regions, farmers aim to limit the negative consequences for production by putting more resources such as capital and fertiliser per unit of output (intensification), and – especially in regions where land is in ample supply (Africa and Latin America) – by converting more land to crop production (extensification). These responses tend to have negative environmental consequences: increased fertiliser use leads to higher emissions and can damage water quality, while conversion of land can have a negative effect on ecosystems and biodiversity, and lead to higher climate change impacts owing to land use and forestry changes. Given also the relatively small share of agriculture in total GDP, the macroeconomic costs of reduced yields as measured by percentage changes in GDP are very limited (-0.1% globally by 2060).

There are significant differences between the regions. Although increases in concentrations of ozone (the driver of the agricultural impacts) will be strongest in China and India (cf. Figure 3.7), the largest projected reductions in agricultural production are in some of the OECD regions, especially USA. The main reason for the strong projected reductions in the USA is that air pollution is already affecting production in the short run, and continues to put downward pressure on agricultural yields and output in the coming decades. In contrast, the consequences for agricultural production in China and India gradually build up over time. Furthermore, there are strong trade links in oilseeds between the USA and Latin America, especially Brazil, so that minor changes in competitiveness between Brazil and USA can translate into relatively large changes in the location of production.

Despite the negative yield impacts in all regions, some regions can increase their crop production beyond the no-feedback level. This effect is strongest in Chile, but also present in e.g. Brazil and the ASEAN economies. Relatively minor domestic yield losses, combined with large opportunities for expanding agricultural land, imply that the *relative* competitive position of these countries in the global crop market is improving vis-à-vis their main competitors. As already mentioned, these economic gains tend to go together with an increased pressure on the environment. This effect mimics similar consequences of yield losses from climate change, and is extensively discussed in OECD (2015) and OECD (2016), which dives into the analysis of how climate change affects trade flows and the revealed comparative advantage of countries.

As indicated in Figure 4.3, large changes in regional agricultural production do not necessarily imply correspondingly large changes in GDP. There are several ways in which GDP is affected by a sectoral shock such as the one on yields. First, in countries where agriculture is a relatively small part of total production, such as the USA, changes in agricultural production do not lead to very significant macroeconomic changes. Second, lower productivity of agriculture also leads to changes in production of other sectors (e.g. the food industry is confronted with higher input costs). Third, the changes in international trade patterns (and terms of trade) resulting from the different changes in agricultural conditions across countries in their agricultural sectors also affect GDP. Fourth, the induced effect of the lower productivity of the economy translates into lower income for households, and thus lower savings. This leads to lower investments and therefore a slow-down of capital accumulation that affects long-term economic growth. However, with a relatively small shock located in one specific sector this effect is limited. These mechanisms illustrate that all sectors and regions are connected to each other, and a shock to one particular part of the economy will have indirect consequences for other regions, other sectors and future time periods.

The largest macroeconomic costs can be found in China, North Africa and India. Furthermore, there are some regions where GDP impacts are negative while consequences for agricultural production are positive, as measured in Figure 4.3 by the total value added in agriculture. A complex set of interactions drives these results, the intensity of which varies between regions. A first explanation is reduced capital accumulation. The lower capital stock hurts all sectors, and especially capital-intensive industries. This is at least partially driven by the fact that these regions are more open to international trade. Hence, a shock to their agricultural system cannot be absorbed domestically, their terms-of-trade deteriorate and their total activity level is lower. Further, some of these regions respond to the agricultural shock by intensifying agricultural production, and when land is abundant, as in the case of e.g. Other Africa, these economies can also resort to extensifying, i.e. increasing agricultural land use. Since agricultural products are necessary goods, with relatively inelastic demand, these regions face negative agricultural shocks by drawing resources away from the rest of the economy. While this may hurt the overall productivity of the economy, it makes sense from the perspective of food security and the basic goods nature of food.

Figure 4.3. **Change in value added and GDP from agricultural impacts, central projection**

Percentage change w.r.t. no-feedback projection, 2060

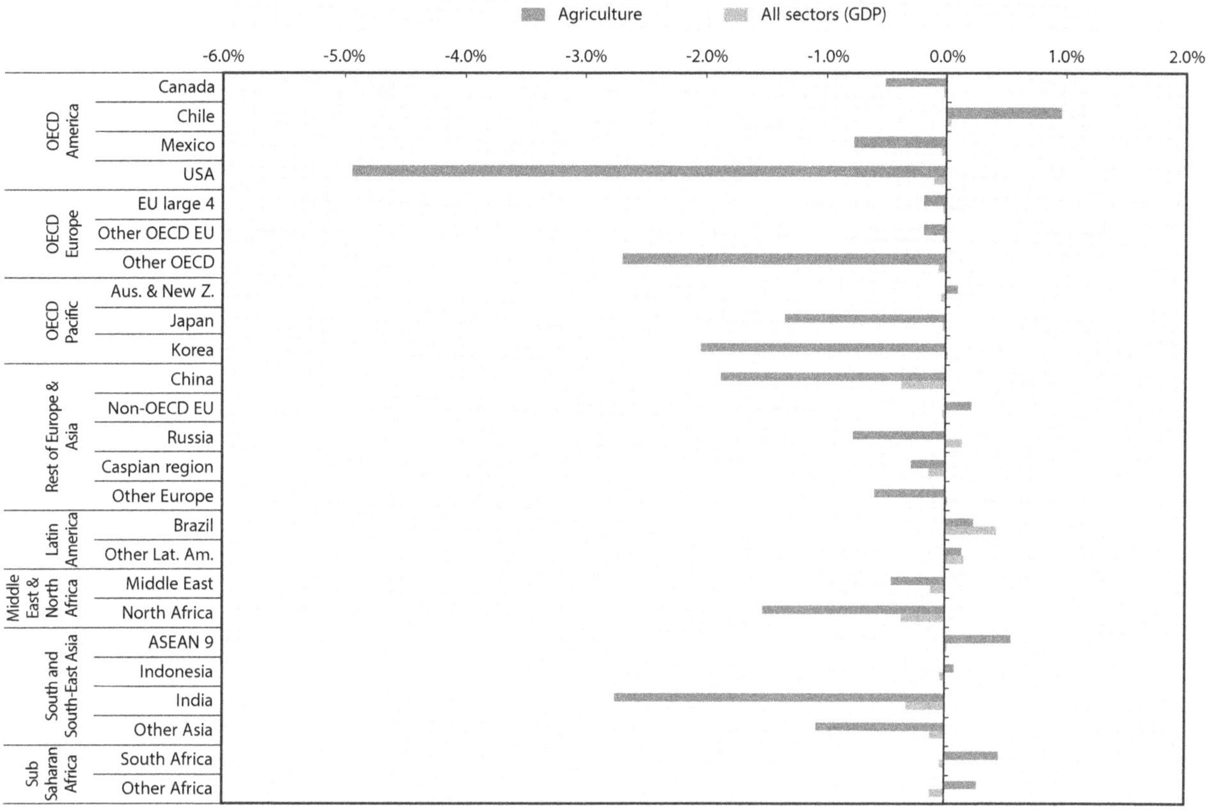

StatLink http://dx.doi.org/10.1787/888933357392

Source: ENV-Linkages model.

4.2. Economic consequences of the combined market impacts

4.2.1. Macroeconomic consequences of the combined market impacts

The three different market impacts of air pollution discussed in Section 4.1 all contribute to a projection of GDP that is below the "naïve" no-feedback projection that excludes the pollution feedbacks on the economy. Panel A in Figure 4.4 summarises how these three impacts evolve over time in terms of percentage changes in global GDP levels from the no-feedback projection. At the global level, the consequences of labour productivity and health expenditure impacts continue to increase significantly relative to GDP. In contrast, agricultural impacts are relatively stable over time in percentage of GDP, i.e. in absolute terms these impacts grow more or less at the same speed as GDP. Taken together, the total annual market costs of outdoor air pollution are projected to rise from 0.3% in 2015 to 1.0% by 2060.

Panel B of Figure 4.4 presents a different way of decomposing the total market costs of outdoor air pollution. The *direct market costs* can be calculated as the sum of the direct economic effects as implemented in the model. This comprises *(i)* the change in value added generated in all sectors from changes in labour productivity; *(ii)* the increased health expenditures; and *(iii)* the change in value added generated in agriculture from changes in crop yields. All these direct costs are measured without taking reallocation of economic resources into account. The *indirect economic effects* can then be deduced as the total macroeconomic costs, i.e. the change in GDP, minus the direct costs. These indirect effects come from reallocation of the factors of production across the economy and e.g. changes in savings rates, and are induced by changes in relative prices. There is a marked difference between the direct and indirect costs: while the direct costs increase more or less at the same pace as economic activity (i.e. the costs in percent of GDP is roughly stable), the indirect costs rapidly increase over time. Two important mechanisms play a key role: *(i)* any negative impact on capital accumulation has a permanent effect as it lowers the growth rate of the economy; and *(ii)* as the shocks become larger over time, the cheapest options are exploited first, and further shocks need to be absorbed at higher costs.

Figure 4.4. **Change in global GDP from combined market impacts, central projection**

Percentage change w.r.t. no-feedback projection

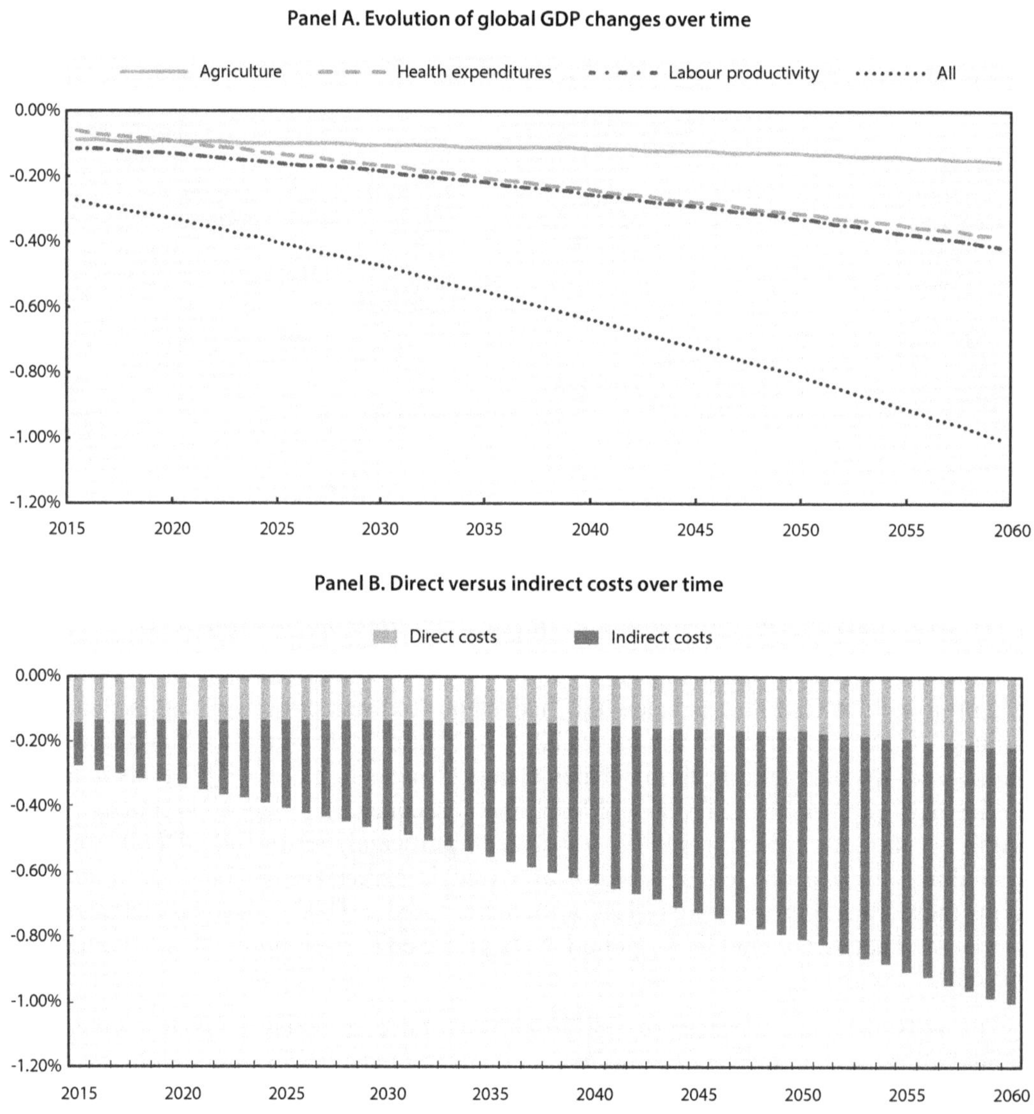

StatLink http://dx.doi.org/10.1787/888933357405

Source: ENV-Linkages model.

At the regional level, it is not surprising that the projected losses are by far the largest in the Rest of Europe and Asia region, which includes China and Russia (Figure 4.5). Not only are the concentrations projected to be very high in this region, the impacts on labour productivity and especially health expenditures are significantly larger than in other regions.[6] The situation is quite different in India. Projected 2060 GDP losses in India are much smaller than in China, despite both countries having projections of very high concentrations (cf. Figures 3.6 and 3.7). One key difference between the two countries is the age structure of the population: India has a much younger population, while aging is projected to become a more severe problem in China. This means that the Chinese population structure in the coming decades is more vulnerable to air pollution, so that for example the additional health expenditures are higher in China than in India. Furthermore, as discussed in Section 4.1, the savings profile of India is significantly different compared with that of China (current savings and investment rates are substantially larger in China, while in the longer run the opposite is true), which imply a different response to a reduction in income or increased expenditures.

Large macroeconomic costs also take place in the Middle East and North Africa and South- and South-East Asia. North Africa is affected by all three market impact categories, while in the Asian regions, one particular impact tends to dominate (labour productivity for India, health expenditures for the ASEAN economies). The projected macroeconomic costs are smaller in the OECD regions, Africa and the Americas.

Figure 4.5. **Change in regional GDP from combined market impacts, central projection**

Percentage change w.r.t. no-feedback projection, 2060

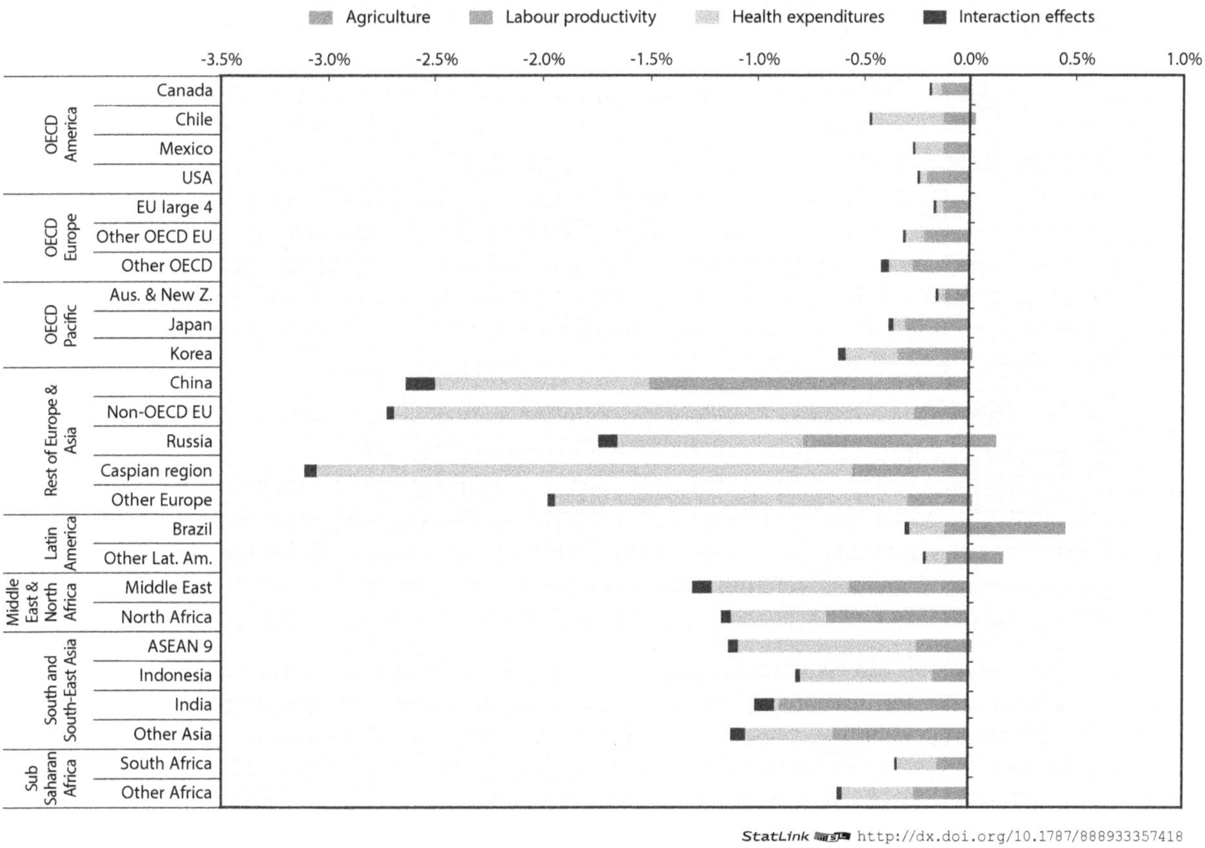

StatLink http://dx.doi.org/10.1787/888933357418

Source: ENV-Linkages model.

The effects of the three different impact categories cannot just be added up to calculate an overall effect of the market impacts of air pollution on economic growth as there are interaction effects that need to be taken into account. In theory, these interaction effects can be both positive and negative. On the one hand, economic consequences tend to be more than proportionally larger for larger shocks, due to the multiplier effect (i.e. that lower income leads to lower savings and thus to lower capital accumulation and lower future income). Thus, combining the different effects may worsen total GDP loss. On the other hand, by combining different shocks into the economic system, a new optimal adjustment process and reallocation of resources may lead to lower costs when combining all impacts. In the projection with all impact categories, the negative effect of having larger distortions of consumption and production possibilities dominates, and the overall GDP loss is larger than the sum of the three individual losses. At the global level this effect is minor (less than 0.1% of GDP in 2060), but for the most affected regions, it can increase GDP losses more significantly.

These effects on economic activity in turn affect emissions of air pollutants. In principle, one should account for these reductions and re-assess the concentration levels and impacts of air pollution until convergence is reached between all steps in the causal chain. However, the reductions in economic activity are fairly limited, and hence emissions levels as projected in the central projection with pollution feedbacks differ less than 1% at the global level from those in the no-feedback projection (and less than 4% at the regional level). Therefore, the second-order effect of lower emission projections on concentrations and impacts is very small, and can be ignored in the light of the uncertainties surrounding all calculations in this report.[7] In other words, there is no need to iterate back from the central projection to revise the no-feedback projection of economic activity and emissions (cf. Section 2.2).

4.2.2. Linking air pollution and climate change

The projected increase in air pollutant emissions also has an effect on climate change. Some air pollutants have a cooling effect (aerosols such as organic carbon), while others are relatively strong near-term climate warmers (esp. black carbon and ozone). To study the interactions between outdoor air pollution and climate change in the projections, radiative forcing have been calculated using the MAGICC6.4 model (Meinshausen et al., 2011). In the no-feedback projection, the aerosols have a direct global *cooling* effect that is projected to increase from 0.4 W/m^2 to 0.5 W/m^2 (excluding indirect effects from induced cloud albedo), while tropospheric ozone has a *warming* effect of similar magnitude. On balance, the contribution of air pollutants to climate change is therefore limited.

The economic feedbacks of outdoor air pollution slow down economic activity around the world, and thus lead to lower global greenhouse gas emissions. However, the effect is fairly minor: less than 1.5% for global emissions, and in all regions less than 4% for carbon dioxide (CO_2), methane (CH_4) and nitrous oxide (N_2O). The resulting reduction in climate impacts is not significant. Reversely, climate damages also have very limited effects on emissions of air pollutants, ranging at the global level from a reduction by 5.5% for NH_3 to an increase of 0.5% for SO_2 according to the projections in OECD (2015).

The interaction effects between climate and air pollution damages may be stronger at the sectoral level, for instance in agriculture. There are also interaction effects on the policy side: reducing the polluting economic activities for air pollution will have significant climate co-benefits. Similarly, mitigation efforts for climate change and air pollution affect emissions of all pollutants; in some cases there are important synergies (e.g. from improvements in energy efficiency), while in other cases trade-offs dominate (e.g. air pollutant capture techniques that reduce the efficiency of power generation). These linkages should be investigated in a comprehensive, integrated manner, but such a study of the multiple benefits of policy action is beyond the scope of this report.

The report *The Economic Consequences of Climate Change* (OECD, 2015) contains a related exercise on the costs of inaction for climate change. It provides a detailed global quantitative assessment of the macroeconomic and sectoral consequences of climate change (i.e. climate damages) for a selected number of impacts: changes in crop yields, loss of land and capital due to sea level rise, changes in fisheries catches, capital damages from hurricanes, labour productivity changes and changes in healthcare expenditures from diseases and heat stress, changes in tourism flows, and changes in energy demand for cooling and heating. It uses the same baseline projection and a very similar production-function methodology.

Figure 4.6. **Outdoor air pollution and climate change impacts, central projection**

Percentage change in GDP w.r.t. no-feedback projection

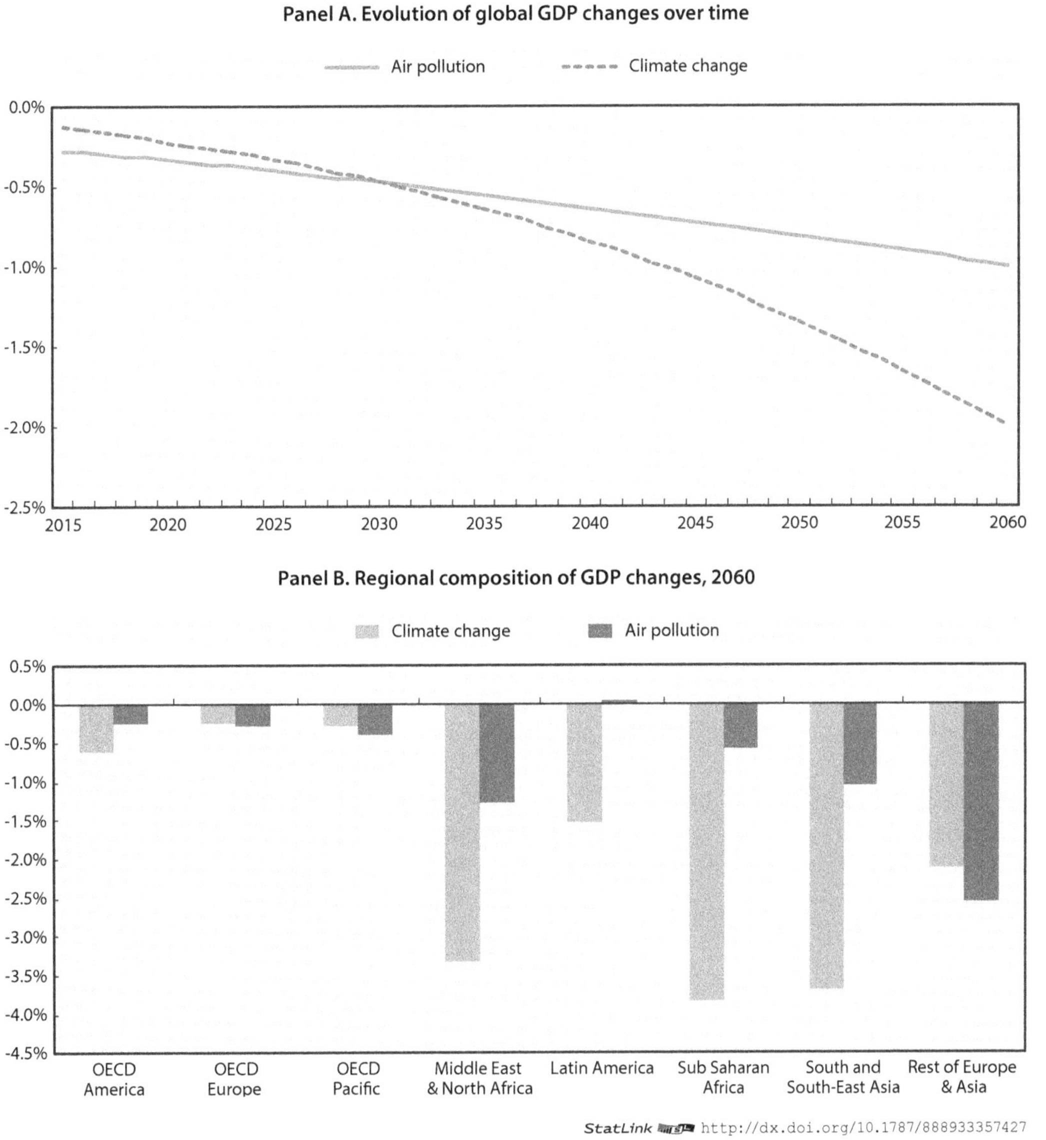

StatLink http://dx.doi.org/10.1787/888933357427

Source: ENV-Linkages model.

There is a much wider literature on the economics of climate change (see OECD, 2015, for an overview). Most directly comparable is the work at JRC-IPTS, that have used similar methodologies to assess the economic consequences of climate change (Ciscar et al., 2011, 2014) and air pollution (Vrontisi et al., 2016).

An important caveat is that, as with the assessment of the economic consequences of air pollution, some of the major consequences do not directly affect markets and could not be accounted for in the modelling framework. The main rationale for policy action on climate change does not come from the market impacts, but rather from the sizable downside risks of tipping points and very severe impacts. Nonetheless, a comparison of the market consequences of climate change and air pollution can help shed light on how these two environmental issues affect economic activity.

In the first half of this century, the order of magnitude of the projected global market costs of air pollution is similar to that of climate change (Figure 4.6). But the time profile of both sets of impacts is very different: air pollution has a stronger effect on the economy in the coming decades, while climate change damages gradually ramp up and become much more significant in the second half of the century. The downside risks of climate change also seem substantially larger, although a proper assessment of the uncertainties surrounding the air pollution damages is not possible owing to a lack of reliable information. Interestingly enough, climate change and air pollution affect the economy through some of the same main channels (labour productivity losses, agricultural yield losses and demand shocks), even if due to different biophysical impacts. But climate change is projected to have more far-reaching macroeconomic consequences and affects a wider set of economic activities, not least capital stocks, directly.

There are also important differences in the geographical distribution of the market costs. OECD (2015) concluded that for climate change even with adaptation "net economic consequences are projected to be [...] especially large in Africa and Asia, where the regional economies are vulnerable to a range of different climate impacts, such as heat stress and crop yield losses". In comparison, the economic consequences of air pollution are much more concentrated in highly populated areas like in Europe and especially in Asia. The position of China and India is also reversed: while climate change impacts are particularly threatening to India, air pollution impacts are larger in China. For geographical reasons, large parts of the OECD are also more affected by air pollution than by climate change, especially in the coming decades.

4.3. Alternative specifications of the market impacts

Applied economic models are based on a series of equations that try to reproduce characteristics of the structure and functioning of the economy. A number of assumptions are needed to set up the modelling frameworks. The modelling assumptions used to model the market impacts of outdoor air pollution reflect the state of the art in the literature (see Vrontisi et al., 2016), but they are still modelling choices and as such influence the results.

This section presents a sensitivity analysis of the results of the market impacts of outdoor air pollution to alternative specifications of the different impacts considered. For the labour market impacts, the central projection only considers the effect of lost working days on labour productivity. Section 4.3.1 presents an alternative specification in which labour supply changes due to premature deaths are also considered. The health expenditure impacts are modelled in the report assuming that households will adjust their consumption levels and that governments will increase their budget to finance the increase in health expenditures through higher taxes on labour. In the alternative specification presented in Section 4.3.2 it is assumed that households and governments will crowd out other expenditures. Finally, Section 4.3.3 presents alternative specification of the impacts of crop yield changes considering the uncertainty ranges relative to the biophysical impacts.

4.3.1. Alternative specification of labour market impacts

The analysis of labour market impacts in the central projection is only based on the direct effect of lost working days on labour productivity and the indirect effects as they emerge in the economy. As an alternative specification, an additional labour supply effect is calculated, by using the premature deaths in the working age population as a shock to labour supply. This additional effect does not aim to resemble a welfare assessment of these premature deaths, but limits itself to identifying the consequences for the economic system through reduced supply of labour. There are several indirect effects that could be taken into account (e.g. lower aggregate consumption due to the decrease in population size or demographic consequences for future generations due to lower births). The net consequence of these effects is not a priori clear and cannot be easily assessed numerically without further examination. Hence, for illustrative purposes, only the direct labour supply effect of the linear projection of premature deaths is included in this alternative specification. The key results are summarised in Figure 4.7.

Figure 4.7. **Sensitivity of market costs to alternative labour market impacts**

Percentage change in GDP w.r.t. no-feedback projection, 2060

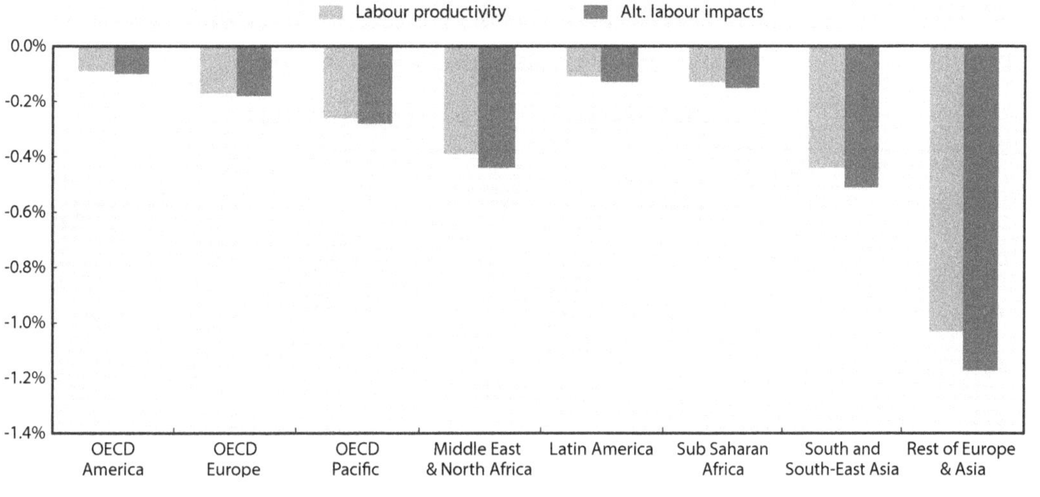

StatLink http://dx.doi.org/10.1787/888933357436

Source: ENV-Linkages model.

While the non-market welfare consequences of premature deaths are very large, the consequences of a smaller labour supply for GDP are minor. In all regions, the consequences of the reduced labour supply are projected to be well below 0.1% of GDP in 2060. This small effect on labour supply and associated GDP losses strengthens the insights from earlier studies (e.g. OECD, 2012) that the key element in an assessment of the economic costs of premature deaths lies in the valuation of the life lost, not in its repercussions in the rest of the economy (see also the discussion in Chapter 5).[8]

4.3.2. Alternative specification of health expenditure impacts

Modelling the response of households and governments to extra health expenditures owing to degradation in health is not straightforward. In the central projection, the assumption is made that households respond by adjusting the consumption levels of other (non-health related) expenditures as well as their savings. Governments are assumed to increase their budget to finance the increase through higher taxes on labour (reflecting the situation in some countries of increased health payments through the social security system). In the alternative specification, the assumption is made that both households and governments respond to the additional health expenditures by fully crowding out expenditures in other commodities. In this crowding out scenario households keep their savings unchanged, while governments keep their budget unchanged.

Figure 4.8 shows how the different regions are affected by the alternative assumption. When health expenditures fully crowd out other expenditures, the consequences of the air pollution impact on GDP levels tend to be significantly smaller. The reasoning is that agents shield the economy from multiplier effects that arise from reducing their savings or increasing their budget. Especially the assumption of fixing private savings to the no-feedback level implies that there is no induced slowdown of the economy through reduced investments and capital accumulation. However, these smaller consequences for GDP do not necessarily imply an improvement in well-being: additional savings come at the expense of consumption, and the government provision of non-health public goods is also reduced. The overall effect of these changes on well-being can unfortunately not be inferred from the modelling framework, which can only measure narrower indicators based on private consumption.

Figure 4.8. **Sensitivity of market costs to alternative health expenditure impacts**

Percentage change w.r.t. no-feedback projection, 2060

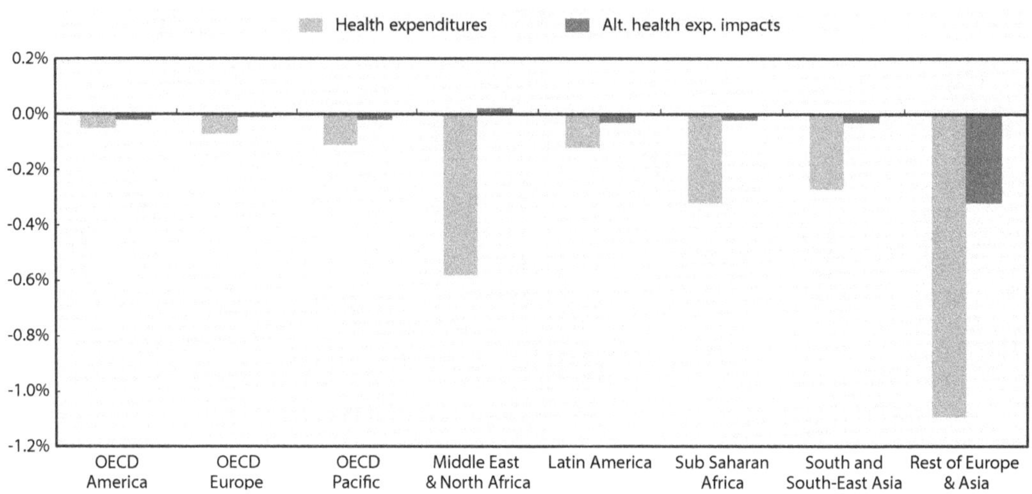

StatLink http://dx.doi.org/10.1787/888933357442

Source: ENV-Linkages model.

4.3.3. Alternative specification of agricultural yield impacts

The calculations of the agricultural impacts are based on the EC-JRC's TM5-FASST model. The model also provides an assessment of the plausible uncertainty range for these impacts, through the calculation of a minimum and maximum impact. These variations are driven by using different metrics for crop varieties and for ozone concentrations (see Section 2.5 and Van Dingenen et al., 2009). The amount of variation between minimum, central projection and maximum varies between crops and regions, but the minimum is roughly half the impact of the central projection, while the maximum is around 50% higher.

Figure 4.9 shows the sensitivity of the economic assessment of the agricultural impacts. The larger the impact, the larger the GDP consequences are, although there are some variations between crops and regions. This pattern extends to the positive consequences in Latin America: larger yield losses in other regions imply more opportunities to increase production in Latin America. Although the domestic negative impacts of air pollution on agricultural production are also larger, what matters more for production in this region is that the difference between its production costs and those of its competitors increase. Hence, their competitive position improves even more when the yield losses are globally larger. This does not imply that larger levels of air pollution are always beneficial for the economies of this region. Once domestic impacts become substantially negative, the negative domestic consequences will start outweighing the increased comparative advantage. Similarly, if global impacts become severe, the slow-down of global demand will also offset the competitiveness gains.

Figure 4.9. **Sensitivity of market costs to alternative agricultural impacts**

Percentage change in GDP w.r.t. no-feedback projection, 2060

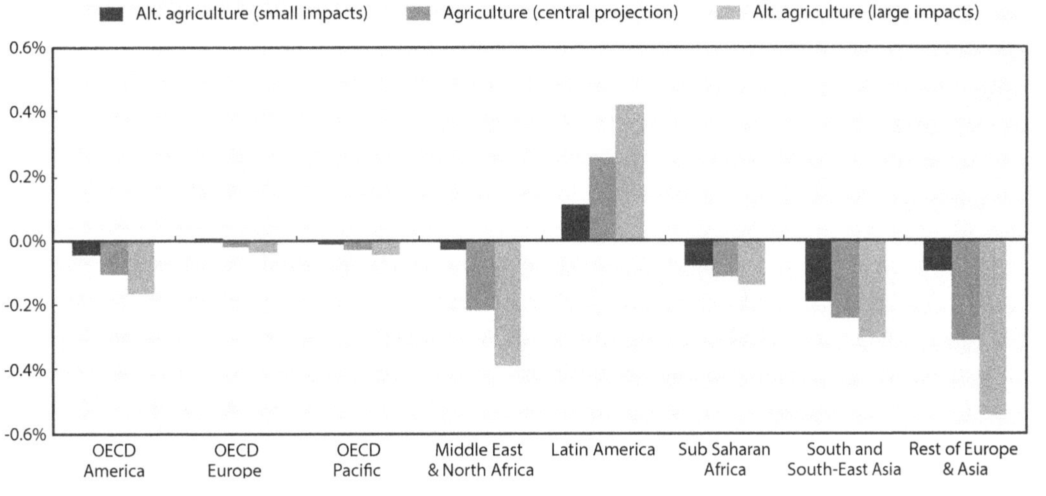

StatLink http://dx.doi.org/10.1787/888933357450

Source: ENV-Linkages model.

Notes

1. The exception is India. Although the labour productivity shock is of the same order of magnitude as that in China, labour represents a significantly larger share of GDP in India. Consequently, the direct labour effect is larger. But it also implies that the reduction in the other components is smaller, and in fact the indirect labour effect turns positive, albeit small.

2. The size of these effects depends on how savings behaviour is modelled, and it does not take into account any specific action of households to change their savings when realising the change in the risk of premature death.

3. Governments could also choose to reduce the quality of the health care that is provided, but the welfare costs of such actions are presumably larger than the health expenditures included here. The costs calculated and presented here are in that sense lower bounds of the potential welfare costs.

4. In all regions, the vast majority of these additional expenditures come from illnesses related to $PM_{2.5}$ concentrations; the contribution of ozone is much smaller.

5. Note that alternative specification of the financing mechanism, e.g. by letting the government budget be balanced by adjusting income taxes or the lump-sum payments between households and government, does not significantly alter these results.

6. Note that these are projections of the costs of policy inaction and do not reflect any judgement on future policy action by China, Russia or any other country.

7. In principle, this does not exclude significant changes for specific hotspots, but the modelling framework does not allow an assessment at that level of detail.

8. However, as this report amply shows, the same does not hold for morbidity costs.

References

Ciscar. J.C. et al. (2014), "Climate impacts in Europe. The JRC PESETA II project", *JRC Scientific and Policy Reports*, No. EUR 26586EN, Publications Office of the European Union, Luxembourg. http://ipts.jrc.ec.europa.eu/publications/pub.cfm?id=7181 (last accessed on 3 May 2016).

Ciscar, J.C. et al. (2011), "Physical and economic consequences of climate change in Europe", *Proceedings of the National Academy of Science (PNAS)*, Vol. 108, pp. 2678-2683.

Meinshausen, M., S.C.B. Raper and T.M.L. Wigley (2011), "Emulating coupled atmosphere-ocean and carbon cycle models with a simpler model, MAGICC6: Part I – Model Description and Calibration", *Atmospheric Chemistry and Physics*, Vol. 11, pp. 1417-1456.

OECD (2016, forthcoming), *The international trade consequences of climate change*, COM/TAD/ENV/JWPTE(2015)63.

OECD (2015), *The Economic Consequences of Climate Change*, OECD Publishing, Paris, http://dx.doi.org/10.1787/9789264235410-en.

OECD (2012), *OECD Environmental Outlook to 2050: The Consequences of Inaction*, OECD Publishing, Paris, http://dx.doi.org/10.1787/9789264122246-en.

Van Dingenen, R. et al. (2009), "The global impact of ozone on agricultural crop yields under current and future air quality legislation", *Atmospheric Environment,* Vol. 43(3), pp. 604-618.

Vrontisi, Z. et al. (2016), "Economic impacts of EU clean air policies assessed in a CGE framework", *Environmental Science and Policy* 55, pp. 54-64.

Chapter 5

Welfare costs of outdoor air pollution to 2060

This chapter presents the results of the analysis of the welfare costs of outdoor air pollution. It starts with an assessment of welfare costs related to the non-market impacts, including both mortality and morbidity, namely those related to the disutility caused by illness. The chapter ends with a discussion of the possibility to compare and add market and non-market costs when they are both expressed as welfare costs. While non-market costs are evaluated through the results of willingness-to-pay studies, market costs are calculated with the ENV-Linkages model and expressed in welfare terms using equivalent variation of income.

5.1. Welfare costs of mortality

As discussed in Chapter 3, air pollution is already the cause of a large number of premature deaths, and pollution-related mortality is projected to increase in the coming decades unless more stringent policies are adopted. It is possible to attribute a cost to these premature deaths with estimates of willingness-to-pay (WTP) based on stated preference (SP) studies. In particular, this report presents the welfare costs of the premature deaths caused by air pollution, calculated using the VSL (see Section 2.6 for an overview of the VSL methodology used).

Table 5.1. **Welfare costs from mortality due to outdoor air pollution, central projection**

Billions of USD, 2010 PPP exchange rates

		2015	2030		2060	
			Non-linear	Linear	Non-linear	Linear
OECD America	Canada	20	30	30	60	60
	Chile	10	10	10	20	20
	Mexico	30	60	60	230	230
	USA	380	460	490	790	830
OECD Europe	EU large 4	360	400	400	500	540
	Other OECD EU	230	310	300	490	490
	Other OECD	140	260	250	670	660
OECD Pacific	Aus. & New Z.	0	10	10	10	20
	Japan	190	270	260	390	400
	Korea	60	130	120	280	290
Rest of Europe & Asia	China	850	2 260	2 450	6 730	8 830
	Non-OECD EU	30	40	40	70	70
	Russia	160	240	240	300	300
	Caspian region	60	150	150	540	560
	Other Europe	30	40	40	90	90
Latin America	Brazil	40	80	80	200	200
	Other Lat. Am.	40	70	80	270	270
Middle East & North Africa	Middle East	80	180	190	770	910
	North Africa	30	60	60	260	270
South and South-East Asia	ASEAN 9	60	140	140	640	750
	Indonesia	30	60	60	230	240
	India	220	570	670	3 360	7 260
	Other Asia	70	150	140	1 070	1 700
Sub-Saharan Africa	South Africa	10	20	20	40	40
	Other Africa	30	50	50	290	300
	World	*3 160*	*6 050*	*6 340*	*18 300*	*25 330*
	OECD	*1 420*	*1 940*	*1 930*	*3 440*	*3 540*
	Non-OECD	*1 740*	*4 110*	*4 410*	*14 860*	*21 790*

StatLink http://dx.doi.org/10.1787/888933357531

Note: Due to the curvature of the functions and rounding, the effects of the non-linear projection can in some cases be reported to be slightly higher than the linear projection; this only affects the results for low and modest concentration levels.

Table 5.1 presents results on the welfare costs associated with the premature deaths caused by outdoor air pollution, relative to both $PM_{2.5}$ and ozone. To facilitate comparison with the modelling results presented in Chapter 4, the national calculations are aggregated into the regional grouping used in the modelling framework. The costs at global level are projected to be close to USD 3.2 trillion in 2015 and increase to USD 18-25 trillion in 2060 (using constant 2010 PPP exchange rates) according to the two different estimates of the number of premature deaths calculated (respectively with linear and non-linear concentration-response function). That is a six- to eightfold increase, which is driven by the increasing number of premature deaths at global level (caused by changes in demographic and concentration trends) and by increasing VSL (following income growth especially in emerging and developing countries).

Welfare costs from premature deaths are by 2060 projected to more than double in OECD countries, going from USD 1.4 trillion in 2015 to USD 3.4-3.5 trillion in 2060. Nevertheless, a larger increase and share of costs are estimated to be in non-OECD economies, where they amount to almost USD 1.7 trillion in 2015 and are projected to increase roughly tenfold to reach USD 15-22 trillion in 2060. That is mostly due to the high number and increase in premature deaths in the People's Republic of China (henceforth "China") and India.

Despite the differences in methodologies, these numbers are comparable to the ones in OECD (2014). OECD (2014) estimates that air pollution caused nearly 500 thousand premature deaths in 2010, corresponding to a welfare cost of USD 1.5 trillion. This report uses the same VSL for OECD countries.

As discussed in Section 2.6, the VSL values used in this report are calculated using a reference OECD value of 2005 USD 3 million and then using benefit transfer techniques to calculate country-specific values following OECD (2012). This is done on the basis of country-specific income and with an income elasticity of 0.8 for high-income countries, 0.9 for middle-income countries and 1 for low-income countries. While this reflects the most reliable values according to the recent literature, there is still a high level of uncertainty surrounding these values.

Figure 5.1 presents a sensitivity analysis on the valuation of premature deaths according to four alternative assumptions on the income elasticities used: *(i)* a uniformly high level, with an elasticity of 1 for all regions; *(ii)* a uniformly low level, with an income elasticity of 0.8 for all regions; *(iii)* a differentiated high level, with 1 for high-income countries, 1.1 for middle-income countries and 1.2 for low-income countries; and *(iv)* a differentiated low level, with 0.6 for high-income countries, 0.7 for middle-income countries and 0.8 for low-income countries.

The figure clearly shows that the uncertainty on the number of deaths (linear versus non-linear) matters more for the assessment of the welfare costs in 2060 than the income elasticity that is used for calculating future values per premature death. The uncertainties are somewhat larger for developing and emerging economies, especially China, than for OECD countries. Including the uncertainty on the valuation broadens the global range of welfare costs from mortality from USD 18.3-25.3 trillion (central projection) to USD 17.2-26.8 trillion. Effectively, at the global level the uncertainty on the valuation increases the uncertainty range of USD 1-1.5 trillion on each side of the range. More than half of that can be attributed to the uncertainty on the values for China. For OECD countries, the uncertainty on the number of deaths tends to be relatively small (cf. Figure 3.8), and the choice of income elasticity matters more. In contrast, for India the uncertainty on the number of deaths is much more important than the uncertainty on the valuation.

Figure 5.1. **Sensitivity of welfare costs from premature deaths to the income elasticity**

Billions of USD, 2010 PPP exchange rates, 2060

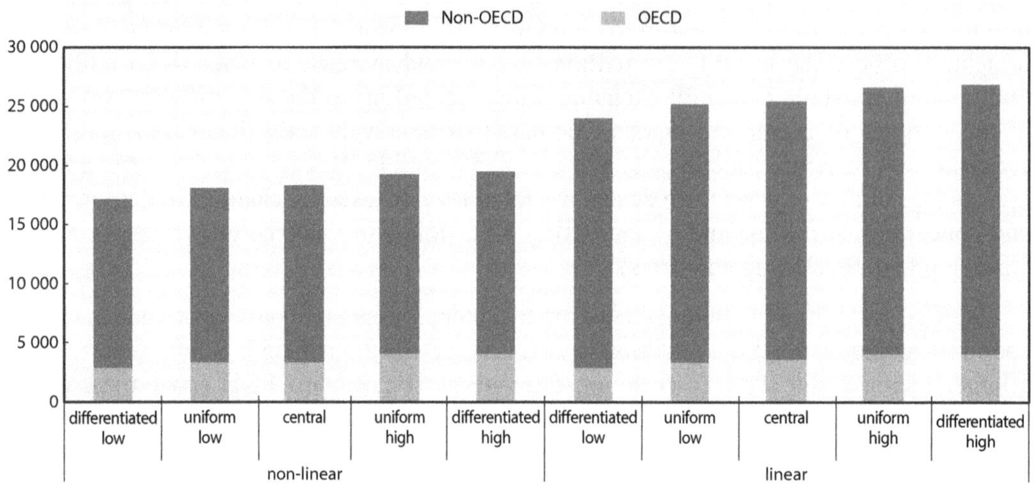

StatLink http://dx.doi.org/10.1787/888933357462

5.2. Welfare costs of morbidity

In this report, the costs related to the disutility of illness are considered to be non-market costs and are estimated using WTP from SP studies, as explained in Section 2.6. Figure 5.2 illustrates the per-capita welfare costs from illness, as broken down into different categories: the costs relative to restricted activity (both restricted and minor restricted activity days), hospital admissions, and illness (new cases of chronic bronchitis in adults, bronchitis in children aged 6 to 12 and asthma symptom days for children aged 5 to 19).

Figure 5.2. **Welfare costs from illness due to outdoor air pollution, central projection**

USD per capita, 2010 PPP exchange rates, 2060

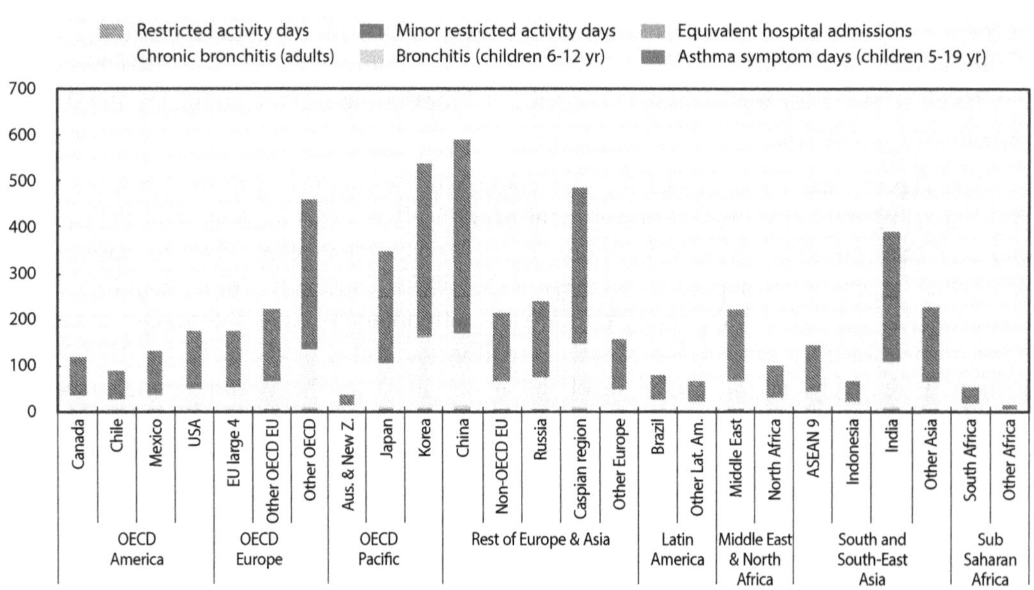

StatLink http://dx.doi.org/10.1787/888933357475

The largest welfare costs come from the restricted activity days, which cause disruptions of normal activities, followed by chronic bronchitis in adults. The regions with the highest per capital costs are China, followed by Korea, Eastern Europe and the Caspian region. These are regions in which the number of cases of illness per capita is highest. Interestingly, Korea and China have similar results, especially for chronic bronchitis in adults. The projected number of cases of chronic bronchitis is higher in China than in Korea (almost 3 million cases in China and 260 thousand cases in Korea in 2060). However, when calculating per capita costs the size of the population matters and it is much higher in China. Further, the value attributed to a single case of adult bronchitis is lower in China than in Korea.

At the global level, welfare costs from non-market impacts of morbidity are estimated to be USD 280 billion in 2015 and USD 2.2 trillion in 2060. This sharp increase over the coming decades shows that an increasing number of people will be affected by air pollution with disruptions to daily life and increasing costs from illness.

5.3. Welfare costs of market impacts

In this report, market costs of health impacts are associated with the effects of the additional health expenditures and changes in labour productivity, as calculated in the general equilibrium model.[1] In addition to these costs, the modelling framework is used to calculate costs of agricultural impacts, plus the indirect costs and effects that take place in the economy, such as sectoral adjustments.

The costs relative to the selected market impacts of air pollution have been presented in Chapter 4 as percentage of GDP. However, GDP cannot directly be compared to the welfare costs of mortality and morbidity presented in Sections 5.1 and 5.2 respectively. Using equivalent variation of income, it is possible to calculate the private welfare costs of the selected impacts of air pollution (excluding welfare losses from the reduced provision of public goods). For more discussion on calculating welfare costs in a computable general equilibrium (CGE) framework, see Section 2.8. GDP and the welfare costs of market impacts are presented in Figure 5.3 as the combined effects of health expenditures, labour productivity, agriculture and an interaction effect.

Figure 5.3. **GDP and welfare costs of market impacts of outdoor air pollution, central projection**

Percentage change of GDP and income w.r.t. no-feedback projection, 2060

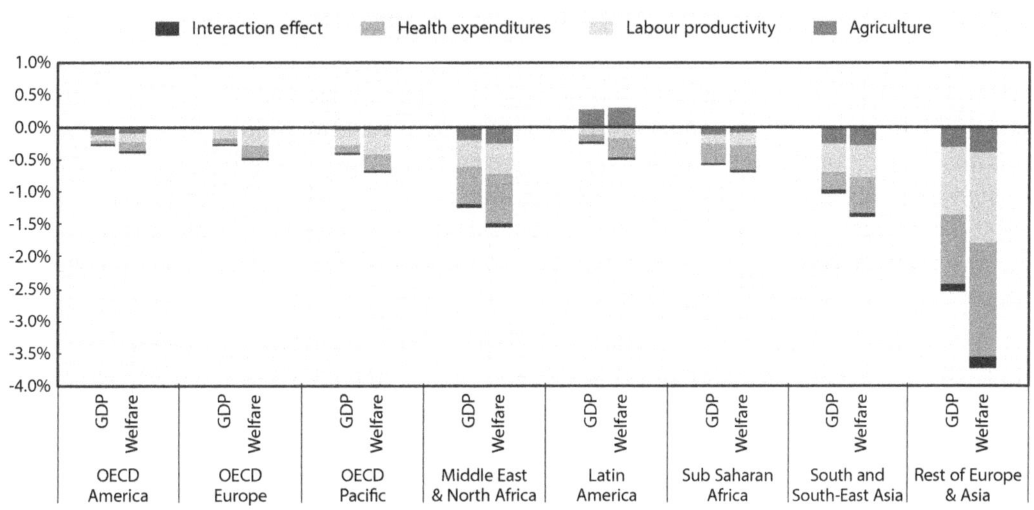

StatLink http://dx.doi.org/10.1787/888933357485

Source: ENV-Linkages model.

The welfare costs are generally larger in percentage change than the GDP impacts: the global cost of air pollution in 2060 is 1.0% of GDP, and 1.5 % of income as calculated with the equivalent variation of income. For agriculture, the equivalent variation is similar to GDP, while for labour productivity and especially health expenditures, welfare costs are larger. The largest difference is for health expenditure, which is the only impact on the demand side. This suggests that impacts on the demand side, which affect private welfare directly, are much larger when considering welfare than when using GDP. The logic for this result is that demand shocks directly affect consumption. The effect of that on welfare cost is corresponding, while it is muted in the change in GDP, which fails to capture the welfare implications of the shocks.

5.4. Bringing together market and non-market costs

Comparing welfare costs from market and non-market impacts

The market costs calculated in the general equilibrium model and expressed in terms of welfare can be compared with the valuation of the non-market welfare costs from premature deaths and disutility from illness. Unfortunately, there is insufficient information to provide an uncertainty range for the costs presented in this section. Only the range for the projected number of premature deaths is included. Therefore, the absolute numbers presented should be treated only as indicative of the order of magnitude of the results, and do not reflect accurate estimates of the welfare costs of outdoor air pollution in the different periods.

Table 5.2 presents the various types of annual welfare costs of air pollution: *(i)* the direct and indirect welfare costs of the selected market impacts of morbidity and agricultural impacts (cf. Section 5.3); *(ii)* the disutility costs from illness (cf. Section 5.2); and *(iii)* the premature deaths due to air pollution (cf. Section 5.1).

The annual welfare costs of the different market impacts in the OECD add up to USD 90 billion in 2015, USD 150 billion by 2030, and USD 390 billion by 2060. That reflects 0.3%, 0.3% and 0.5% of income (as measured in GDP per capita), respectively; or USD 70, USD 110 and USD 270 per capita. At the global level, the numbers are larger, both in absolute terms and as percentage of income, and rising much more rapidly over time: while in 2015 and 2030 the average welfare costs of the market impacts per person are lower in non-OECD economies than in the OECD region, by 2060 they are substantially higher in non-OECD economies, reaching 1.5% of income.

For the OECD as a whole, the annual welfare costs related to non-market health impacts of outdoor air pollution amount to up to USD 1.6 trillion by 2015, and rise to USD 3.9 trillion in 2060, of which more than 90% stem from the welfare loss of premature deaths. At the global level, the costs are projected to be USD 3.4 billion in 2015 and are rising more rapidly, reaching USD 6.6-6.9 trillion by 2030, and USD 20.5-27.6 trillion by 2060. This larger uncertainty band reflects the sensitivity of the projected premature deaths at very high concentration levels, where the concentration-response function potentially becomes non-linear (see Section 5.1).

These welfare costs from non-market impacts are not related to expenditures or tradable goods; they can therefore not be directly compared with macroeconomic indicators such as GDP. But to give a sense of the order of magnitude of these welfare costs, one can express them as a share of total income; for the OECD countries combined this is around 5% in 2015, and remain roughly constant over time. At the global level, they increase from 6% in 2015 to 9-12% in 2060.

Table 5.2. **Total welfare costs of outdoor air pollution, central projection**

Billions of USD, 2010 PPP exchange rates

	OECD			World		
	2015	**2030**	**2060**	**2015**	**2030**	**2060**
Welfare costs from market impacts						
Agriculture						
Direct costs	10	10	20	40	50	80
Indirect economic effects	10	20	40	50	90	320
Health: Morbidity						
Health expenditures						
Direct costs	10	10	30	20	40	140
Indirect economic effects	20	40	100	120	290	1 350
Labour productivity						
Direct costs	30	40	60	50	90	350
Indirect economic effects	10	30	120	30	140	900
Economic interaction effects	0	0	20	20	30	160
TOTAL market impacts	**90**	**150**	**390**	**330**	**730**	**3 300**
Share of income (percentage)	*0.3%*	*0.3%*	*0.5%*	*0.6%*	*0.7%*	*1.5%*
Per capita (USD per capita)	*70*	*110*	*270*	*50*	*90*	*330*
Welfare costs from non-market impacts						
Health						
Morbidity: Disutility costs	130	170	310	280	560	2 240
Mortality	1 420	1 930-1 940	3 440-3 540	3 160	6 050-6 340	18 300-25 330
TOTAL non-market impacts	**1 550**	**2 100-2 110**	**3 750-3 850**	**3 440**	**6 610-6 900**	**20 540-27 570**
*Share of income (percentage)**	*5%*	*4%*	*5%*	*6%*	*7%*	*9-12%*
Per capita (USD per capita)	*1 210*	*1 530-1 540*	*2 610-2 680*	*470*	*780-820*	*2 060-2 770*
Other costs						
Missing effects (biodiversity, cultural heritage, ...)	N/A	N/A	N/A	N/A	N/A	N/A

StatLink http://dx.doi.org/10.1787/888933357541

* Welfare costs from non-market impacts are not related to expenditures and therefore not an integral part of the calculation of income; the expression of these welfare costs as share of income is therefore only for illustrative purposes.

Finally, one can represent these non-market welfare costs also in per capita terms. In 2015, the per-capita welfare costs of outdoor air pollution for non-market impacts are higher in OECD countries than in the emerging and developing countries: around USD 1 200 per capita for the OECD, and less than USD 500 per capita for the world. By 2060, the situation is changed, despite continued population growth in developing countries: per capita costs in the OECD region are projected to rise modestly to USD 2 610-2 680, whereas they increase to USD 2 060-2 770 globally. A large share of increasing non-OECD costs takes place in the Rest of Europe and Asia region (incl. China), as previously discussed. This reflects both the high concentration levels and the increase in costs associated to the health impacts that follows economic growth and rising income levels.

While it is clear that by far the largest cost component is the welfare loss from premature deaths, indirect economic consequences as induced by the various market impacts have an increasingly important role. When using welfare as a measure for the market costs, indirect economic effects are calculated as the difference between direct market costs and the equivalent variation of income.

In the short- and medium term, indirect economic repercussions tend to be of the same order of magnitude as the direct market impacts. But in the long run (2060), the induced economic consequences of air pollution will outweigh the direct effects of the various market impacts, not least due to the long-term consequences of a slowdown of economic growth. Ignoring these indirect economic consequences can lead to a significant miscalculation of the morbidity costs of air pollution. Figure 5.4 confirms the increasing importance of the indirect economic consequences over time.

Figure 5.4. **Evolution of the welfare costs of outdoor air pollution over time, central projection**

Billions of USD, 2010 PPP exchange rates

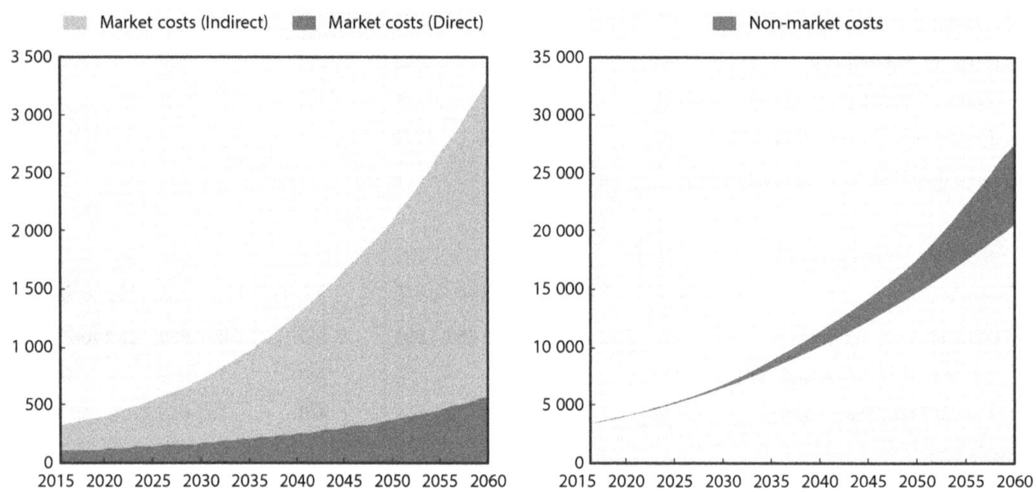

StatLink http://dx.doi.org/10.1787/888933357495

Aggregating welfare costs from market and non-market impacts

The total welfare costs of the impacts of outdoor air pollution comprise both market and non-market costs. In principle, market and non-market costs should be added up, as each part only paints a partial picture of the total welfare costs. However, this is rarely done in the literature because studies generally have focused on only one dimension of the total welfare costs. For example, the valuation literature mostly focuses on non-market costs, and ignores indirect economic effects (Hunt et al., 2016). On the other hand, the cost-of-illness and CGE modelling literature can calculate direct and indirect economic effects, but generally cannot deal with non-market impacts (e.g. Vrontisi et al., 2016).

The advantage of the comprehensive approach taken in this report is that it provides detailed projections of both market and non-market costs. Complications can arise with aggregating the two, as measurement techniques differ and as it is impossible to ensure that all possible sources of double-counting are excluded.

The welfare costs of market and non-market impacts calculated in this report are measured differently but can both be expressed as aggregate income losses. On the one hand, in the CGE modelling assessment of market costs, the equivalent variation of income reflects the maximum willingness to pay to avoid the deterioration in the economic system resulting from the market impacts of outdoor air pollution. This assessment assumes that households behave rationally and focuses purely on changes in private consumption. On the other hand, the valuation of non-market costs is based on studies directly asking respondents to value a change in risk. Relying on available estimates from the literature for non-market values implies that there is a potential that the underlying questionnaires are not fully compatible with the rest of the assessment presented in this report. This needs to be kept in mind when interpreting the aggregate results.

There is also a risk that certain costs are double counted. As explained in Section 2.6, double-counting is avoided as much as possible. Agricultural impacts are assessed only as market costs, and non-market costs are ignored.[2] Similarly, for mortality effects, double counting is excluded by focusing purely on the non-market costs, as these likely dominate and the valuation of mortality reflects total values. For morbidity effects, double counting is potentially a more significant problem because both market and non-market costs are considered, but the unit values used for disutility are based on studies that, at least in principle, cover only non-market costs and exclude all market costs (OECD, 2012).

Notwithstanding these difficulties, it is legitimate to assess the full cost of inaction by summing the monetary values of the different cost components, provided the caveats are kept in mind. The uncertainties described above mean that the absolute numbers presented in this section should be interpreted with care. It is not the point estimate of the costs of inaction itself but the order of magnitude that should incentivise policy action. The numbers on the total welfare costs of outdoor air pollution presented could be seen as an upper bound of the full welfare costs related to the impacts considered given the potential for double-counting. However, these welfare costs exclude certain impacts that are likely to have negative consequences for welfare, such as the direct health effects of NO_2 or the effects on ecosystems and biodiversity, which imply that potential total welfare costs of outdoor air pollution are likely higher than those presented in this report.

Summing the different cost elements presented in Table 5.2, the total global welfare costs of outdoor air pollution from all impacts that could be measured in this report are projected to be around USD 3.8 trillion (7% of income; USD 510 per capita) in 2015, and rising to USD 23.8-30.9 trillion (11-14% of income; USD 2 400-3 100 per capita) by 2060. In comparison, the corresponding total welfare costs for the OECD region amount to USD 1.6 trillion (5% of income; USD 1 280 per capita) for 2015 and USD 4.1-4.2 trillion (5% of income; USD 2 880-2 950 per capita) for 2060, respectively.

Regional differences are especially strong for the indirect economic effects, as Figure 5.5 illustrates (using linear values for mortality). As discussed in detail in Chapter 4, in some regions, such as Eastern Europe, the indirect effects of health expenditures are especially strong and negative, and substantially worsen the welfare consequences of air pollution. But for other regions, the indirect economic consequences are much more benign, as countries can increase their competitive position relative to their competitors. This is for example the case for Brazil and other Latin American countries in the agricultural sector. This reduces the negative economic consequences, and could potentially even lead to absolute gains in economic activity, and hence GDP and welfare.

Figure 5.5. **Components of regional welfare costs of outdoor air pollution, central projection**

Shares in total welfare costs based on linear values for mortality, 2060; numbers in brackets represent the share of mortality costs for the range of linear and non-linear values

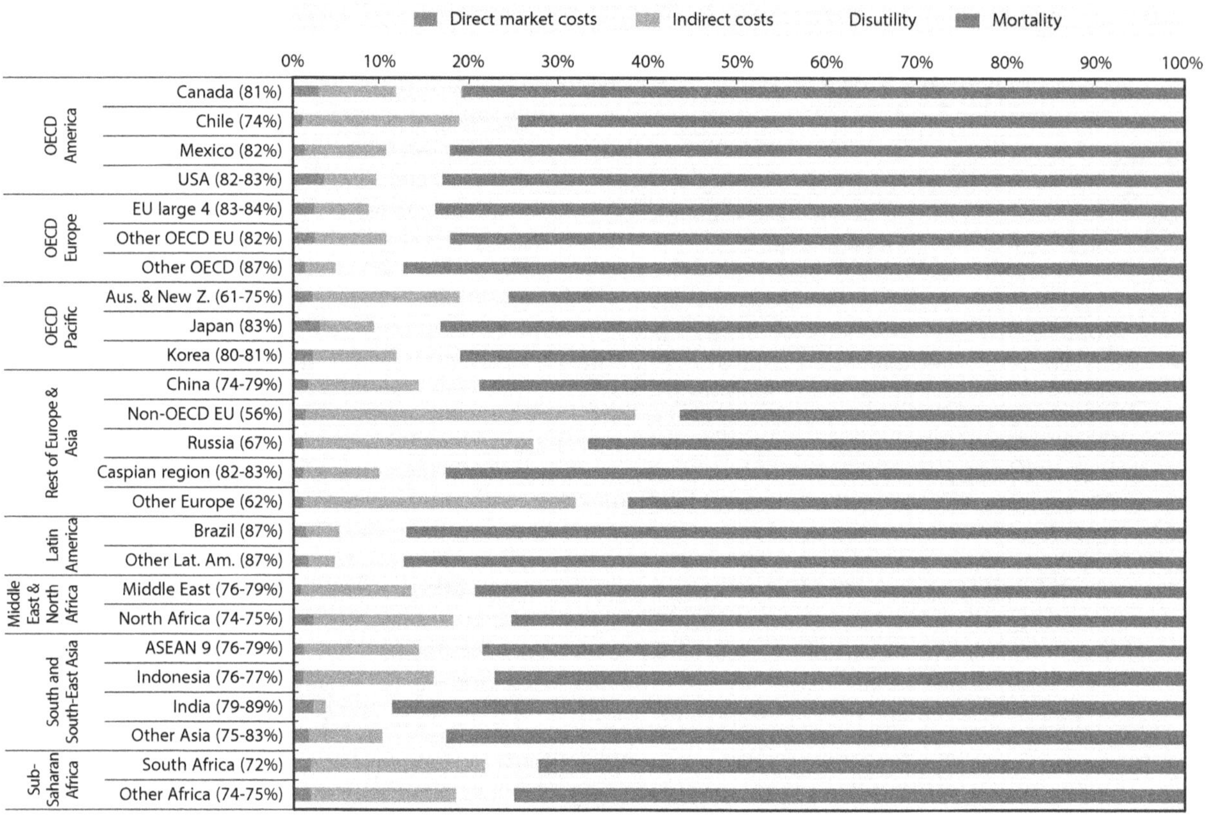

StatLink http://dx.doi.org/10.1787/888933357502

As the share of the indirect effects increases, total morbidity costs are projected to also grow more rapidly than the costs of mortality.[3] OECD (2014) and Hunt et al. (2016) suggest using a 10% mark-up on mortality costs as a proxy for morbidity costs, based on earlier valuation studies. The results presented in this report, with all their caveats, roughly confirm that such a ratio seems adequate for short term global assessments when the indirect economic effects are small. In fact, the ratio roughly holds globally throughout the model horizon when indirect economic effects are ignored. However, this mark-up should increase over time, as in the longer run indirect economic effects are stronger. Furthermore, a generic mark-up ignores the significant differences between regions.

These results regarding the importance of indirect effects support the need to study both market and non-market impacts for the assessment of the full costs of morbidity, and hence the full costs of outdoor air pollution. This can be done with a combination of suitable tools for different types of costs, including an economic systems model for the (indirect) market costs, and direct valuation of non-market welfare costs based on WTP from SP studies.

Notes

1. Costs related to loss of leisure time could not be captured in the modelling framework.
2. The main reason for this is that there is insufficient data at global level to adequately quantify the welfare costs of e.g. ecosystem and biodiversity losses that are associated with agricultural impacts.
3. There also other factors that help explain the lower share of mortality in the long run, including improvements in health care that may avert deaths.

References

Hunt, A. et al. (2016), "Social costs of morbidity impacts of air pollution", *OECD Environment Working Papers* No. 99, OECD Publishing, Paris, http://dx.doi.org/10.1787/5jm55j7cq0lv-en.

OECD (2014), *The Cost of Air Pollution: Health Impacts of Road Transport*, OECD Publishing, Paris, http://dx.doi.org/10.1787/9789264210448-en.

OECD (2012), *Mortality Risk Valuation in Environment, Health and Transport Policies*, OECD Publishing, Paris, http://dx.doi.org/10.1787/9789264130807-en.

Annex A

Description of the ENV-Linkages model

The OECD's in-house dynamic CGE model – ENV-Linkages – is used as the basis for the assessment of the economic consequences of climate impacts until 2060. The advantage of using a CGE framework to model climate impacts is that the sectoral details of the model can be exploited. Contrary to aggregated IAMs, where monetised impacts are directly subtracted from GDP, in a CGE model the various types of impacts can be modelled as directly linked to the relevant sectors and economic activities.

ENV-Linkages is a multi-sectoral, multi-regional model that links economic activities to energy and environmental issues. The ENV-Linkages model is the successor to the OECD GREEN model for environmental studies (Burniaux, et al. 1992). A more comprehensive model description is given in Chateau et al. (2014); whereas a description of the baseline construction is given in Chateau et al. (2011).

Production in ENV-Linkages is assumed to operate under cost minimisation with perfect markets and constant return to scale technology. The production technology is specified as nested constant elasticity of substitution (CES) production functions in a branching hierarchy (cf. Figure A.1). This structure is replicated for each output, while the parameterisation of the CES functions may differ across sectors. The nesting of the production function for the agricultural sectors is further re-arranged to reflect substitution between intensification (e.g. more fertiliser use) and extensification (more land use) of crop production; or between intensive and extensive livestock production. The structure of electricity production assumes that a representative electricity producer maximises its profit by using the different available technologies to generate electricity using a CES specification with a large degree of substitution. The structure of non-fossil electricity technologies is similar to that of other sectors, except for a top nest combining a sector-specific resource with a sub-nest of all other inputs. This specification acts as a capacity constraint on the supply of the electricity technologies.

The model adopts a putty/semi-putty technology specification, where substitution possibilities among factors are assumed to be higher with new vintage capital than with old vintage capital. In the short run this ensures inertia in the economic system, with limited possibilities to substitute away from more expensive inputs, but in the longer run this implies relatively smooth adjustment of quantities to price changes. Capital accumulation is modelled as in the traditional Solow/Swan neo-classical growth model.

The energy bundle is of particular interest for analysis of climate change issues. Energy is a composite of fossil fuels and electricity. In turn, fossil fuel is a composite of coal and a bundle of the "other fossil fuels". At the lowest nest, the composite "other fossil fuels" commodity consists of crude oil, refined oil products and natural gas. The values of the substitution elasticities are chosen as to imply a higher degree of substitution among the other fuels than with electricity and coal.

Figure A.1. **Production structure of a generic sector in ENV-Linkages**

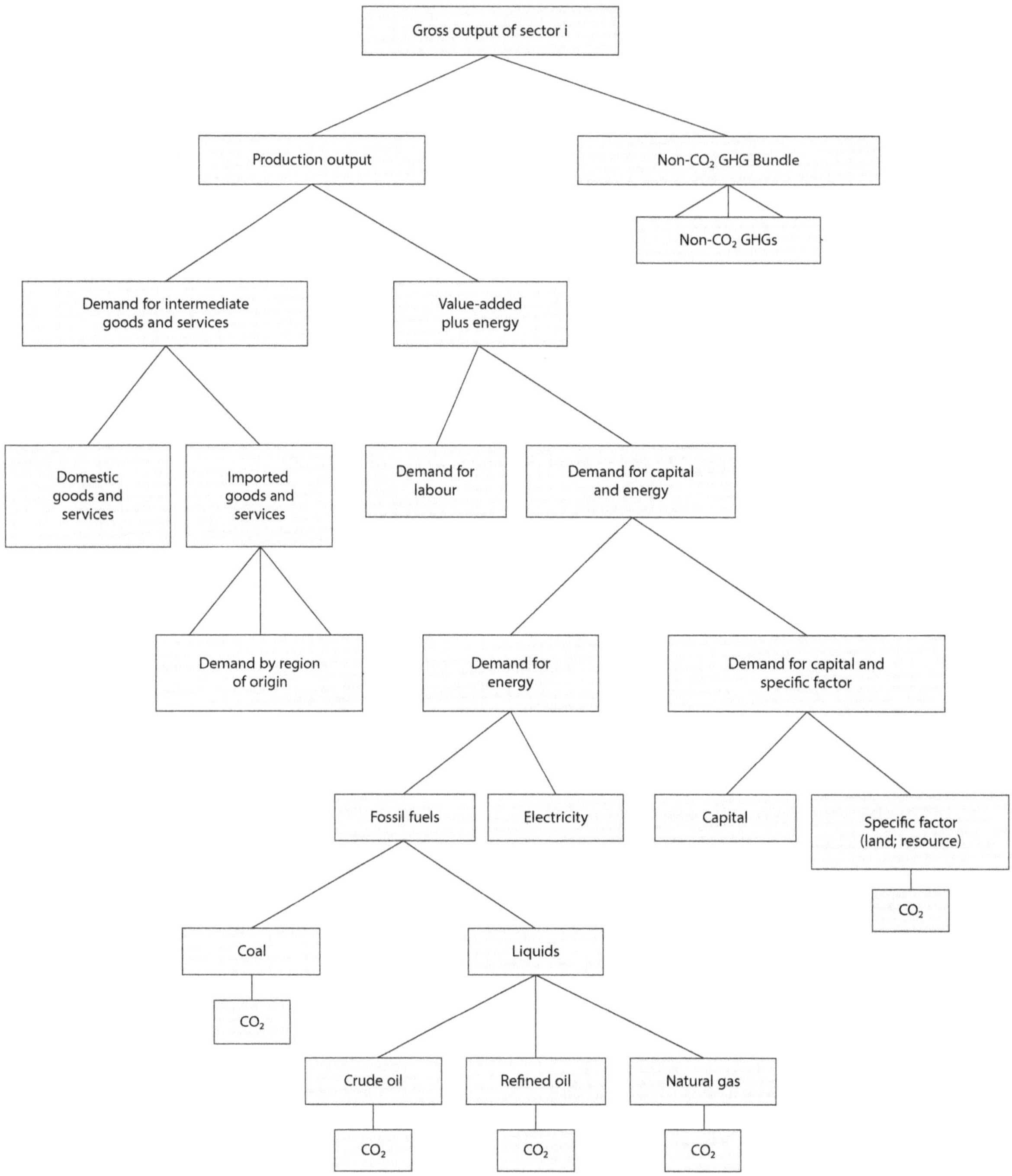

Household consumption demand is the result of static maximisation behaviour which is formally implemented as an "extended linear expenditure system". A representative consumer in each region – who takes prices as given – optimally allocates disposal income among the full set of consumption commodities and savings. Saving is considered as a standard good in the utility function and does not rely on forward-looking behaviour by the consumer. The government in each region collects various kinds of taxes in order to finance government expenditures. Assuming fixed public savings (or deficits), the government budget is balanced through the adjustment of the income tax on consumer income. In each period, investment net-of-economic depreciation is equal to the sum of government savings, consumer savings and net capital flows from abroad.

International trade is based on a set of regional bilateral flows. The model adopts the Armington specification, assuming that domestic and imported products are not perfectly substitutable. Moreover, total imports are also imperfectly substitutable between regions of origin. Allocation of trade between partners then responds to relative prices at the equilibrium.

Market goods equilibria imply that, on the one side, the total production of any good or service is equal to the demand addressed to domestic producers plus exports; and, on the other side, the total demand is allocated between the demands (both final and intermediary) addressed to domestic producers and the import demand.

CO_2 emissions from combustion of energy are directly linked to the use of different fuels in production. Other greenhouse gas (GHG) emissions are linked to output in a way similar to Hyman et al. (2002). The following non-CO_2 emission sources are considered: *(i)* methane from rice cultivation, livestock production (enteric fermentation and manure management), fugitive methane emissions from coal mining, crude oil extraction, natural gas and services (landfills and water sewage); *(ii)* nitrous oxide from crops (nitrogenous fertilizers), livestock (manure management), chemicals (non-combustion industrial processes) and services (landfills); *(iii)* industrial gases (SF_6, PFCs and HFCs) from chemicals industry (foams, adipic acid, solvents), aluminium, magnesium and semi-conductors production. Over time, there is, however, some relative decoupling of emissions from the underlying economic activity through autonomous technical progress, implying that emissions grow less rapidly than economic activity.

Emissions can be abated through three channels: *(i)* reductions in emission intensity of economic activity; *(ii)* changes in structure of the associated sectors away from the "dirty" input to cleaner inputs, and *(iii)* changes in economic structure away from relatively emission-intensive sectors to cleaner sectors. The first channel, which is not available for emissions from combustion of fossil fuels, entails end-of-pipe measures that reduce emissions per unit of the relevant input. The second channel includes for instance substitution from fossil fuels to renewable in electricity production, or investing in more energy-efficient machinery (which is represented through higher capital inputs but lower energy inputs in production). An example of the third channel is a substitution from consumption of energy-intensive industrial goods to services. In the model, the choice between these three channels is endogenous and driven by the price on emissions.

ENV-Linkages is fully homogeneous in prices and only relative prices matter. All prices are expressed relative to the *numéraire* of the price system that is arbitrarily chosen as the index of OECD manufacturing exports prices. Each region runs a current account balance, which is fixed in terms of the *numéraire*. One important implication from this assumption in the context of this report is that real exchange rates immediately adjust to restore current account balance when countries start exporting/importing emission permits.

As ENV-Linkages is recursive-dynamic and does not incorporate forward-looking behaviour, price-induced changes in innovation patterns are not represented in the model. The model does, however, entail technological progress through an annual adjustment of the various productivity parameters in the model, including e.g. autonomous energy efficiency and labour productivity improvements. Furthermore, as production with new capital has a relatively large degree of flexibility in choice of inputs, existing technologies can diffuse to other firms. Thus, within the CGE framework, firms choose the least-cost combination of inputs, given the existing state of technology. The capital vintage structure also ensures that such flexibilities are larger in the long-run than in the short run.

The sectoral and regional aggregation of the model, as used in the analysis for this report, are given in Tables A.1 and A.2, respectively.

Table A.1. **Sectoral aggregation of ENV-Linkages**

Agriculture	Manufacturing
Paddy rice	Paper and paper products
Wheat and meslin	Chemicals
Other grains	Non-metallic minerals
Vegetables and fruits	Metals n.e.s.
Sugar cane and sugar beet	Fabricated metal products
Oil seeds	Other manufacturing
Plant fibres	Motor vehicles
Other crops	Electronic equipment
Livestock	Textiles
Forestry	
Fisheries	
Natural resources and energy	**Services**
Coal	Land transport
Crude oil	Air transport
Gas extraction and distribution	Water transport
Other mining	Construction
Petroleum and coal products	Trade other services and dwellings
Electricity (5 technologies*)	Other services (government)

* Fossil fuel based electricity: combustible renewable and waste based electricity; nuclear electricity; hydro and geothermal; solar and wind.

Table A.2. **Regional aggregation of ENV-Linkages**

Macro regions	ENV-Linkages countries and regions
OECD America	Canada Chile Mexico United States
OECD Europe	EU large 4 (France, Germany, Italy, United Kingdom) Other OECD EU (other OECD EU countries) Other OECD (Iceland, Norway, Switzerland, Turkey, Israel)
OECD Pacific	Oceania (Australia, New Zealand) Japan Korea
Rest of Europe and Asia	People's Republic of China Non-OECD EU (non-OECD EU countries) Russia Caspian region Other Europe (non-OECD, non-EU European countries)
Latin America	Brazil Other Lat.Am. (other Latin-American countries)
Middle East & North Africa	Middle-East North Africa
South and South-East Asia	India Indonesia ASEAN9 (other ASEAN countries) Other Asia (other developing Asian countries)
Sub-Saharan Africa	South Africa Other Africa (other African countries)

References

Chateau, J., C. Rebolledo and R. Dellink (2011), "An Economic Projection to 2050: The OECD "ENV-Linkages' Model Baseline", *OECD Environment Working Papers*, No. 41, OECD Publishing, Paris, http://dx.doi.org/10.1787/5kg0ndkjvfhf-en.

Chateau, J., R. Dellink and E. Lanzi (2014), "An Overview of the OECD ENV-Linkages Model: Version 3", *OECD Environment Working Papers*, No. 65, OECD Publishing, Paris, http://dx.doi.org/10.1787/5jz2qck2b2vd-en.

Annex B

Description of the TM5-FASST model

TM5-FASST is a global air quality source-receptor model (AQ-SRM), developed by the European Commission's Joint Research Centre in order to address the need for swift and easy evaluation of global and regional air pollution emission scenarios and their impacts on human health and ecosystems. In general, AQ-SRMs link emissions of pollutants in a given source region with downwind impacts, using knowledge of meteorology and atmospheric chemical and physical processes which transform the emitted pollutant precursors. The source region is any point or area from which emissions are considered; the receptor is any point or area at which the pollutant concentration and impact is to be evaluated. An AQ-SRM will then include a functional relation between each emitted precursor and each end product for each source region and each receptor region.

The TM5-FASST model is a reduced-form SRM: the relation between the emissions of compound *i* from source *x* and resulting pollutant *j* concentration (where $j = i$ in case of a primary component) at receptor *y* is expressed by a simple functional relation, which mimics the underlying meteorological and chemical processes. In the current version of TM5-FASST, the function is a simple linear relation:

$$C_{ij}\left(x,y\right)=C_j^0\left(y\right)+A_{ij}\left(x,y\right)\cdot E_i\left(x\right)$$

where $C_{ij}(x, y)$ is the concentration of species *j* at receptor *y* formed from precursor *i* emitted at source *x*, $E_i(x)$ is the emission rate (kg/yr) of precursor *i* at source *x*, $A_{ij}(x, y)$ is the so-called source-receptor coefficient (SRC) between source location *x* and receptor location *y* for emitted precursor *i* leading to end product *j*, and $C_j^0(y)$ is a constant for pollutant *j* and location *y*.

The SRCs have been derived from a set of runs with the full chemical transport model TM5-CTM (Krol et al., 2005) by applying emission perturbations for each of a defined set of source regions and precursor components. TM5-CTM explicitly solves the mass balance equations of the species using detailed meteorological fields and sophisticated physical and chemical process schemes. TM5-CTM covers the global domain with a resolution of 1°×1°. More in particular, the applied procedure to calculate the SRCs was based on 56 source regions covering the global continents.

A base run with a reference global emission dataset for all relevant pollutants and pollutants precursors for the year 2000 was performed, including SO_2, NO_x, BC, OC, NMVOC, and NH_3. This run is based on the IPCC AR5 RCP reference scenario for the year 2000 (Van Vuuren et al., 2012). The base run produces the resulting base concentrations of all relevant pollutants at a global 1°×1° resolution.

A series of perturbation runs was performed, where sequentially in each of the defined 56 source regions, the emission of each of the pollutant precursors was reduced over the entire source region by 20% relative to the base run, and the resulting concentration of all affected pollutant species was calculated, in the same way as it was done for the base run. Hence, in principle, the number of perturbation runs is 56×n, with n the number of emitted compounds considered to be relevant. In practice, in order to reduce the number of runs, some non-interacting compounds were grouped into one perturbation simulation. For CO which is a longer-lived species perturbations were made at the aggregation level of continents. For CH_4 a single global perturbation run with TM5-CTM from the HTAP1 modelling experiment was used to evaluate the response on background ozone per kg emitted CH_4 (Fiore et al., 2009). The difference between the concentration field for a specific compound from each perturbation run and the base run is a global 360×180 concentration field (1°×1° resolution), the so-called delta-field.

For each receptor point (each grid cell), the resulting delta concentration between base and perturbation run, leads to the calculation of a unique SRC, expressing the concentration response in each grid cell upon an emission change in source region x as in the following equation, where $\Delta E_i(x) = 0.2 \times E_i^0(x)$ with $E_i^0(x)$ the base run emission.

$$A_{ij}(x,y) = \frac{\Delta C_j(y)}{\Delta E_i(x)}$$

Hence, the total concentration of component j in receptor region y, resulting from arbitrary emissions of all its precursors i at all source regions x is obtained by scaling the respective SRCs with the actual emission changes:

$$C_{ij}(x,y) = C_j^0(y) + \sum_x \sum_i A_{ij}(x,y)\left[E_i(x) - E_i^0(x)\right]$$

For example, in the case of j=ozone, the i precursors would comprise NO_x, NMVOC, CO and CH_4. An overview of all considered precursor-pollutant combinations is given in Table B.1. This set of linear equations for all components and all source and receptor regions emulates the full-fledged TM5-CTM, and constitutes the "kernel" of TM5-FASST.

Table B.1. **Relevant emitted precursor-pollutant pairs in TM5-FASST**

Pollutant →	SO_2	NO_x	NH_3	O_3	CH_4	SO_4	NO_3	NH_4	BC	POM	SO_x	NOY	BC
Precursor ↓	gas	gas	gas	gas	gas	PM	PM	PM	PM	PM	dep	dep	dep
SO_2 (g)	XXX	X	XX	X	X	XXX	XX	XX			XXX		
NO_x (g)	X	XXX	XX	XXX	XX	XX	XXX	XX			X	XXX	
NH_3 (g)	X	X	XXX	X	X	XX	XX	XXX			X		
BC (g)									XXX				XXX
POM (g)										XXX			
NMVOC (g)	X	X	X	XXX	XX	X	X	X			X		
CO (g)				XXX	XX								
CH_4 (g)	X	X	X	XXX	XXX	X	X	X			X		

Note: The number of x's gives a qualitative indication of the most influential precursors (xxx: highest influence). Influences indicated by a single x are due to feedback mechanisms affecting the level of oxidants, and hence the lifetime of hydroxyl radical (OH), in the atmosphere, which in turn affects the oxidation rate of the precursors. The (g) refers to gaseous component; PM = particulate matter; dep = deposited component. POM = polycyclic organic matter; NO_Y: fixed nitrogen.

Source: TM5-FASST model.

The resulting global concentration maps for different emission scenarios obtained by applying the source-receptor coefficients provide the required information to further assess the impact of emissions changes in terms of effects on human health, vegetation and ecosystems in general.

A full description of the TM5-FASST model methodology and validation against the full TM5 model is given by Van Dingenen and Dentener (forthcoming).

References

Fiore, A. et al. (2009), "Multimodel estimates of intercontinental source-receptor relationships for ozone pollution", *Journal of Geophysical Research*, 114, D04301.

Krol, M. et al. (2005), "The two-way nested global chemistry-transport zoom model TM5: algorithm and application", *Atmospheric Chemistry and Physics,* Vol. 4, pp. 3975-4018.

Van Dingenen, R. and F.J. Dentener (forthcoming), "TM5-FASST: a global source-receptor model for particulate matter, trace gases and radiative forcing", manuscript in preparation for the special HTAP issue in ACP.

Van Vuuren, D.P. et al. (2012), "A proposal for a new scenario framework to support research and assessment in different climate research communities", *Global Environmental Change*, Vol. 22(1), pp. 21-35.

Annex C

Methodology to calculate the health impacts

Following the quantification of population exposure to air pollution using the TM5-FASST model, analysis of health impacts proceeds by combining information on concentration response functions, the fraction of population at risk and the incidence of ill-health, to quantify the health impacts of air pollution, including e.g. the number of cases of mortality, hospital admissions, and chronic bronchitis.

Mortality

The Global Burden of Disease (GBD) mortality results for 2010 from the work of Forouzanfar et al. (2015) and Brauer et al. (2016) for $PM_{2.5}$ impacts and Lim et al. (2012) for ozone impacts are taken as the starting point for all countries, together with annual average population-weighted exposure data at the national level for $PM_{2.5}$, and the mean of 6-month maximum concentration (M6M) for ozone. The GBD results were adopted as they were derived from a major international peer-reviewed exercise, carried out at a higher level of spatial disaggregation than was possible in this study. Given that the GBD estimates are limited to the present time and based on cause-specific analysis of mortality, it is necessary to consider the extent to which changes in health to 2060 will affect the results. An analysis of WHO data as related to UN-sourced population data carried out for this study found that observed changes in the cause of death in each region over time are not so large as to add significant uncertainty to the analysis.

For ozone, a linear model was adopted where a unit change in M6M generated the same change in risk throughout the concentration range generated for the study in excess of a counterfactual concentration adopted by Lim et al. (2012). Changes in projected mortality rates for future years were also factored into the analysis using data from the UN's World Population Projections (UN, 2012).

GBD has adopted a non-linear response function for quantification of the effects of $PM_{2.5}$, an approximation of which has been implemented here. The non-linearity in the curve is intended to account for an expected decline in response per unit of exposure as concentrations rise. Noting uncertainty in the development of this function (there is little information available to inform the shape of the relationship at high ambient concentrations typical of those in countries like the People's Republic of China and India where the majority of impacts are expected to occur), a linear function has also been derived for $PM_{2.5}$ for the present study. For the non-linear relationships, the mortality estimates for $PM_{2.5}$ from GBD were analysed to parameterise the following equations, generating individual estimates of α and β for each country:

$$PAF_{GBD} = \begin{cases} \alpha \ln(concentration\ in\ \mu g / m^3) + \beta, & for\ concentrations > 5.8\ \mu g / m^3 \\ 0, & otherwise \end{cases}$$

Where *PAF* = pollution attributable fraction, and α and β are curve-fitting coefficients to the GBD results. Impacts on mortality are therefore measured against a reference level of pollution (5.8 mg/m^3) below which it is assumed that the impacts of outdoor air pollution on health do not lead to any premature deaths. The value 5.8 µg/m^3 is a counterfactual or "cut-off" concentration below which no additional health impacts are calculated. It is not a health risk threshold, as emerging epidemiological evidence finds adverse health burdens for even lower concentrations (e.g. Shi et al., 2016).

The number of premature deaths is then calculated based on the *PAF*, following this equation:

$$Deaths = PAF_{GBD} \times CMR \times Population$$

Where *CMR* = crude mortality rate. CMR is taken from the World Bank's World Development Indicators (World Bank, 2015) for 2010, and the UN's World Population Prospects (UN, 2012) for subsequent years under the median fertility projection. For year 2010, the deaths calculated with the specified equation match the number of deaths calculated by the GBD study.

The alternative linearised model for quantifying $PM_{2.5}$ impacts on mortality was derived in a similar way to the model used for ozone, again accounting for changes in mortality rates in future years. Together these relationships provide a range for mortality impacts with the non-linear function providing the lower projection and the linear function the upper projection.

Morbidity

For analysis of morbidity (illness) impacts, the analysis is based on the conclusions of the HRAPIE (Health Response to Air Pollutants in Europe) study (WHO, 2013), which was used in the cost-benefit analysis of the European Commission's Clean Air Policy Package of December 2013 (Holland, 2014; European Commission, 2013). It is acknowledged that other groups have developed or applied alternative sets of response functions for morbidity, including USEPA (2011, for the prospective analysis of the benefits of the US Clean Air Act to 2020). The HRAPIE conclusions were adopted here because the study, led by WHO, is both recent and involved experts from a large number of countries in both Europe and North America. The effects quantified using the HRAPIE functions were as follows:

- $PM_{2.5}$
  - Effects of chronic (long term) exposure on adult and childhood bronchitis;
  - Effects of acute (short-term) exposure on hospital admissions for respiratory and cardiovascular illness, restricted activity days, lost working days, asthma symptom days for children;
- Ozone
  - Effects of acute (short-term) exposure on hospital admissions and "minor" restricted activity days.

Bronchitis takes a different course for adults and children. For adults, the disease, once initiated, is long lasting, often persisting until death, and varying in severity from minor to severe. For children, however, the disease is short-lived, lasting for about 2 weeks on average. These differences are reflected in the economic valuation. Although there are more cases of childhood bronchitis, the longer lasting cases of adult bronchitis generate larger economic damage.

Quantification of these morbidity effects requires knowledge of incidence rates across the population. Whilst these data are available for a growing number of countries they are not available for all. This problem has been identified in previous work carried out for OECD on transport (OECD, 2014), with morbidity costs quantified as a fixed proportion of mortality costs, 10%, referenced against cost-benefit analyses for the European Commission and US-EPA. An advantage of this approach is that it automatically factors in the question of non-linearity in response functions in a manner that it is consistent with the approach taken for mortality.

To provide analysis for all countries it is therefore necessary to extrapolate results. The approach taken here is broadly similar to that used in OECD (2014) but more detailed. Results from the analysis of the European Commission's Clean Air Policy Package, for which the HRAPIE functions had been applied in full, were adopted as the basis for this extrapolation. It was assumed that there would be a linear relationship between mortality and morbidity. In theory, higher rates of mortality might reduce the population at risk of bronchitis and other illnesses. The position taken here assumes that air pollution related mortality does not significantly affect the population at risk as exposure levels rise. Using results from Holland (2014) averaged ratios between mortality and morbidity across 28 countries from the European results were obtained. These were then adjusted to account for differences in mortality estimates for European countries between Holland (2014) and GBD.

For ozone, a single estimate was made for each morbidity effect, whilst for $PM_{2.5}$, two estimates, linked to the linear (upper projection) and non-linear (lower projection) mortality functions were derived for all years after 2010. Results for 2060 are shown in Table C.1. Chapters 4 and 5 illustrate results only relative to the upper estimates (which for 2060 are roughly 50% greater than the lower projections). Preference for the upper projection for morbidity can be justified from the perspective that only a subset of possible impacts can be quantified at the present time (RCP, 2016, provides a commentary on the variety of impacts that can be linked to air pollution over the life course).

Table C.1. **Range of health impacts at global level for 2060**

Respiratory diseases (million number of cases)	
Bronchitis in children aged 6 to 12	24-36
Chronic bronchitis (adults, cases)	7-10
Asthma symptom days (million number of days)	
Asthma symptom days (children aged 5 to 19)	230-360
Healthcare costs (million number of admissions)	
Hospital admissions	8-11
Restricted activity days (million number of days)	
Lost working days	2 460-3 750
Restricted activity days	9 820-14 900
Minor restricted activity days (asthma symptom days)	2 580

It is acknowledged that the extrapolation of morbidity results for Europe to the rest of the world is subject to a number of uncertainties, most importantly that:

- Due to lack of data, it assumes similar prevalence rates for each disease throughout the world to those seen in European countries, when these will of course vary substantially. It does not recognise variation in rates of specific diseases in the same way that the GBD analysis does for cause-specific mortality.
- It implicitly assumes that healthcare provision is similar in all countries, when it patently is not. Hence for hospital admissions, it implies that European admission rates are typical of all other countries, when there is substantial variation around the world with respect to access to healthcare systems. The problem is most serious for extrapolation for developing countries as this is where the majority of impacts are expected to occur. Most of these countries will have a lower level of healthcare provision than is typical of European countries. Against this, however, the lack of healthcare facilities clearly does not mean that everyone is well. People will still experience the illness. Indeed, it may become significantly worse through the lack of healthcare, with the result that the impact is more severe than it would otherwise be. Thus, while this assumption may imply an overestimation of health expenditures, it underestimates welfare costs.
- A similar issue arises with respect to lost working days. The European results are based on European rates of absenteeism, with a certain standard of social welfare and employment conditions that is not universal. Employees without these conditions may be inclined to go to work when they would otherwise be considered unwell ("presenteeism"), and may take longer to recover (or alternatively, develop worse illness), and therefore also have lower labour productivity.

Valuation

The approach to valuation is described in Chapter 2, with the unit values used for analysis shown in Table 2.2. These values are adjusted to account for economic conditions in each country, and the development of the economies over time.

Three elements are considered for morbidity valuation, healthcare costs, lost productivity, and welfare losses through pain, suffering, etc. There is some potential in doing this for double counting of costs, for example, if willingness-to-pay estimates account implicitly for lost productivity and healthcare costs. However, Table C.2 demonstrates that such potential double-counting has been avoided largely by attributing endpoints to a specific cost component.

Table C.2. **Cost components to the valuation of health endpoint**

	Welfare cost	Healthcare cost	Productivity cost
Deaths	100%	0%	0%
Chronic bronchitis (adults, cases)	82%	18%	0%
Bronchitis in children aged 6 to 12	92%	8%	0%
Equivalent hospital admissions	14%	86%	0%
Restricted activity days (all ages)	100%	0%	0%
Minor restricted activity days (children 5-19 yr)	100%	0%	0%
Lost working days	0%	0%	100%

Only three impacts (bronchitis in adults, bronchitis in children, equivalent hospital admissions) are assessed under more than one of the categories applied. All three effects combine welfare cost with healthcare cost, and in each case, one of the value categories dominates with more than 80% of total value (hence the maximum extent of any double counting for these effects is of the order of 25%). In selecting the welfare valuation data the main sources used have been European, reducing the probability that respondents would have included healthcare costs in their response given typical European models for funding healthcare. Combined with the view (RCP, 2016) that there are a number of impacts that could be added to the analysis, it is concluded that double counting of health related costs is insignificant to the analysis.

References

European Commission (2013), "The Clean Air Policy Package", *Commission Staff Working Document. Impact Assessment,* No. SWD(2013)531, European Commission, Brussels, http://ec.europa.eu/governance/impact/ia_carried_out/docs/ia_2013/swd_2013_0531_en.pdf (last accessed on 3 May 2016).

Forouzanfar, M.H. et al. (2015), "Global, regional, and national comparative risk assessment of 79 behavioural, environmental and occupational, and metabolic risks or clusters of risks in 188 countries, 1990-2013: a systematic analysis for the Global Burden of Disease Study 2013", *The Lancet*, Vol. 386(10010), pp. 2287-2323.

Holland, M. (2014), Cost-benefit analysis of final policy scenarios for the EU Clean Air Package. Corresponding to IIASA TSAP report no.11, Version 2a. Report to European Commission DG Environment. March 2014, http://ec.europa.eu/environment/air/pdf/TSAP%20CBA.pdf (last accessed on 3 May 2016).

Lim, S.S. et al. (2012), "A comparative risk assessment of burden of disease and injury attributable to 67 risk factors and risk factor clusters in 21 regions, 1990-2010: A systematic analysis for the Global Burden of Disease Study 2010", *The Lancet,* Vol. 380, pp. 2224-60.

OECD (2014), *The Cost of Air Pollution: Health Impacts of Road Transport*, OECD Publishing, Paris, http://dx.doi.org/10.1787/9789264210448-en.

RCP (Royal College of Physicians) (2016), *Every breath we take: the lifelong impact of air pollution.* Report of a working party of the Royal College of Physicians. London, www.rcplondon.ac.uk/projects/outputs/every-breath-we-take-lifelong-impact-air-pollution (last accessed on 3 May 2016).

UN (United Nations) (2012), *World Population Prospects, the 2012 Revision*, http://esa.un.org/unpd/wpp/index.htm (last accessed on 3 May 2016).

WHO (World Health Organisation) (2013), *Health risks of air pollution in Europe – HRAPIE project. Recommendations for concentration-response functions for cost-benefit analysis of particulate matter, ozone and nitrogen dioxide*, World Health Organization, Regional Office for Europe, Bonn, Germany, www.euro.who.int/en/health-topics/environment-and-health/air-quality/publications/2013/health-risks-of-air-pollution-in-europe-hrapie-project-recommendations-for-concentrationresponse-functions-for-costbenefit-analysis-of-particulate-matter,-ozone-and-nitrogen-dioxide (last accessed on 3 May 2016).

ORGANISATION FOR ECONOMIC CO-OPERATION AND DEVELOPMENT

The OECD is a unique forum where governments work together to address the economic, social and environmental challenges of globalisation. The OECD is also at the forefront of efforts to understand and to help governments respond to new developments and concerns, such as corporate governance, the information economy and the challenges of an ageing population. The Organisation provides a setting where governments can compare policy experiences, seek answers to common problems, identify good practice and work to co-ordinate domestic and international policies.

The OECD member countries are: Australia, Austria, Belgium, Canada, Chile, the Czech Republic, Denmark, Estonia, Finland, France, Germany, Greece, Hungary, Iceland, Ireland, Israel, Italy, Japan, Korea, Luxembourg, Mexico, the Netherlands, New Zealand, Norway, Poland, Portugal, the Slovak Republic, Slovenia, Spain, Sweden, Switzerland, Turkey, the United Kingdom and the United States. The European Union takes part in the work of the OECD.

OECD Publishing disseminates widely the results of the Organisation's statistics gathering and research on economic, social and environmental issues, as well as the conventions, guidelines and standards agreed by its members.

OECD PUBLISHING, 2, rue André-Pascal, 75775 PARIS CEDEX 16
(97 2016 09 1 P) ISBN 978-92-64-25746-7 – 2016

www.ingramcontent.com/pod-product-compliance
Lightning Source LLC
Chambersburg PA
CBHW061222240926
51846CB00006B/1313